DESIGN ROADMAPPING

Lianne Simonse

BIS PUBLISHERS

Colophon

AUTHOR
Lianne Simonse

ART DIRECTOR
Barbara Iwanicka

TEXT EDITOR
Jianne Whelton

DR. IR. LIANNE W.L. SIMONSE
Faculty of Industrial Design Engineering
Delft University of Technology
Landbergstraat 15
2628 CE Delft
The Netherlands
T +31 (0)15 27 89054
L.W.L.Simonse@tudelft.nl

BIS PUBLISHERS
Building Het Sieraad
Postjesweg 1
1057 DT Amsterdam
The Netherlands
T +31 (0)20 515 02 30
bis@bispublishers.com
www.bispublishers.com

B/SPUBLISHERS

ISBN 978-90-6369-459-3

Copyright © 2017 Lianne Simonse
and BIS Publishers.
Amsterdam, The Netherlands.

All rights reserved. No part of this publication may be reproduced or transmitted in any form or by any means, electronic or mechanical, including photocopy, recording or any information storage and retrieval system, without permission in writing from the copyright owners.

Every reasonable attempt has been made to identify owners of copyright. Any errors or omissions brought to the publisher's attention will be corrected in subsequent editions.

Foreword

DESIGN ROADMAPPING is for anyone interested in design, strategy and innovations and its wonderful combination. Design roadmapping is a process that enables organisations and designers to devise creative responses to future strategic challenges. This book is written for designers, strategists and innovators alike, to guide on your journey of roadmapping the future with the land marks of products and services you envisioned together with a team of innovators.

A journey that can start with uncovering new trends, scouting for new technologies and mapping the values on the roadmap. Along the way, you can map a future vision, frame the time pacing and create the pathways towards it. To arrive at the end of the roadmapping process with the result of a clear, complete and compelling design roadmap. Such an artifact can provide your organisation strong visualisation and decision support for your organisation's future plans on design innovations.

In this book you will find guidelines on how to roadmap altogether with explorations and definitions of theoretical concepts. Besides deep insights on the origins, theories and science results, designers, strategists and innovators share how-to examples and cases on their roadmapping experiences and achievements.

Writing this book has been a truly fascinating journey for me. Not only to express my passion for the subject that has captivated me for several years now, but also to be able to share this in a visual compelling way. Creating a design book comes with the joy of visualising the story as much as the freedom of writing it.

This exploration of design roadmapping stems directly from my working life as an industrial consultant, roadmapper, researcher and teacher in a design school. Because of these experiences, the question of the relationship between innovations and future time has emerged for me not only as an intellectual puzzle but also as an object of personal quest. I have become convinced that roadmapping is not just the business alignment of the long term strategy of technology research and engineering, with the strategic channelling of new inventions into breakthrough applications, parallel to the re-engineering of existing products into more cost effective solutions.

In my experience, at the heart of roadmapping are the user values that drive the future timing of innovation. Only when users embrace a new innovation, when there is a close connection between their value wishes, desires and needs, and only then when the time has come that a critical mass agrees on the value of the innovation, it can become successful. Therefore, in design roadmapping, future visions are built on value drivers that connect user values to an evolutionary pacing of design innovations. Mapping value innovations by the pacing of time intervals ensures continuous innovation for your organisation on the long run.

Design roadmapping is devoted to the values of users and committed to a time-based view of design innovations, design roadmapping fosters strategic design to practical competence and professional artistary. Although, not as a job of a lone designer, but as a team endeavour. Design roadmapping is a process in which creative conversations and multiple sessions build the common ground for the future plans of innovation.

Guidebook structure

This book covers the creation of a design roadmap from an initial effort on value mapping, to idea mapping, onwards to pathway mapping. We provide guidelines to create a compelling, visionary and actionable roadmap. Seven topics are unraveled in seven chapters. On each topic a case, a science interview and a lab for active self-study is provided.

We start with a general introduction on DESIGN ROADMAPPING. In this first chapter we will define what a roadmap is, describe the overall process of roadmapping and also take a deep-dive into the history of roadmapping and grasp some inspiration from cartographic roadmaps. We further used the roadmapping process (see figure 1.3) to structure this guidebook. Each chapter addresses one of the key activities from the process of design roadmapping.

The first key activity of the value mapping stage: CREATIVE TREND RESEARCH is the central topic of the second chapter. We discuss a creative fusion between visual trend spotting and strategic trend scanning, and present four different trend techniques. Followed by the second key activity in the value mapping stage of FUTURE VISIONING. In the third chapter we look closely at what a future vision is and what it is not. We discuss artifacts that exemplify future visions such as concept cars and we explore the leadership position in future visioning.

The next two key activities of the idea mapping stage: TECHNOLOGY SCOUTING and TIME PACING STRATEGY are the subject of the following two chapters. The fourth chapter takes a look at the heritage of technology roadmapping, and offers updated guidelines for the design research of technology scouting. The fifth chapter details how to strategically time the pace of the organisaton's innovations. To help you properly time the pace, we provide you with important information regarding possible transitions in strategic lifecycles and for different types of design innovations.

The sixth chapter takes us to the topic of organising MAPPING SESSIONS. Elementary in the process of roadmapping, is to talk about the future, to share imaginations, gut feelings and beliefs; and to create a mutual understanding of the promising opportunities, the current positioning, the future vision and the various possible ideas and pathways towards it. The sixth chapter further elaborates on how to organise a chain of mapping sessions to carry out the roadmapping proces. The final chapter is about how to VISUALISE A ROADMAP and how to tailor visualisations to suit different stages of the roadmap process and different kinds of audiences.

Acknowledgement

In creating this book I have been extremely lucky to be able to work with many creative people, designers, strategist and innovators remarkably skilled to conceive wonderful, original ideas for valuable experiences in the future and amazing roadmaps. It is absolutely fabulous to work with top designers and - scientist from the field who helped me shaping the conceptual framing of the book. I am especially grateful to Flavio Manzoni, Enrico Leonardo Fagone, Paul Hilkens, Bart Massee, Sabina Popin, Sasha Abram, Bianca Mediati, Dirk Snelders, Deborah Nas, Ena Voûte, Susan Reid, Gloria Barczak, Pascal Le Masson, Erik-Jan Hultink, Helen Perks and Roberto Verganti.

It has been rewarding to meet so many talented students that took inspiration from the design roadmapping master course. I am especially thankful to those students that warmly welcomed the request to provide their roadmap artifacts. Many thanks Esmee Mankers, Pepijn van Dalen, Luuk Roos, Zoë Dankfort, Tess Poot, Tonino Gatti, Lars Scholings, Eva van Genuchten, Pepijn van der Zanden, Mark Kwanten, Nienke Nijholt, Ben Hup, Joana Portnoy, Anne Brus, Ruben Verbaan, Marco Bonari, Wouter Aerts, Niya Stoimenova, Eva Frese, Niels Corsten, Robin Kwa, Robert Stuursma, Zhang Ziyi, Yee Jek Khaw and Roël Tibosch. This book would not have been possible without you all devoting time and energy to visualise and document the wonderful and inspiring roadmaps. I want to take the opportunity to thank you all.

For making the design book a visual and enjoyable read. I want to thank Barbara Iwanicka for sharing her great expertise on the secrets of graphic book design. Also thanks for the enthousiasm and positive feedback from the artists and photographers who provided their inspirational work full off reflections that have kept our spirits high during this captivating project. Thanks Fiona Tan, Maarten Baas, Julie Merethu, Christopher Prentiss Michel, Ari Versluis en Ellie Uyttenbroek, Daan Roosegaarde, Ogre Bot and Jonathan Harris. I want to thank Jianne Whelton, our English editor for her devotion to make the book an enjoyable read and Bionda Dias for all her efforts on the organisation and the production of the book. The support and faith of my friends, family members and colleagues, has been vital. I deeply treasure the loving support of Willie Koehorst for being my sparring partner throughout the whole project and providing me with insights and fruitful comments on the book's raw version. Finally, a project involving such a large group of collaborators would not have been realized without the financial support provided by the faculty of Industrial Design Engineering. I want especially thank Daniëlle van der Kruk, who runs our department of product innovation management in an amazing way.

There are certainly many roadmaps and endeavors that have not found their way to the pages of this book. Nonetheless they all are important in contributing to the body of knowledge on design roadmapping that continues to grow. Let's continue building even better future visions and roadmaps! It has been a truly fascinating experience for me. I hope you will enjoy, as I did, and experience some inspiration to create your own design roadmaps.

Lianne Simonse

Contents

3   Foreword
6   Contents

8   **DESIGN ROADMAPPING**
    10   Design Roadmap
    16   Process of Design Roadmapping
    19   Roadmapping Performance
    26   Metaphors-Case
    30   Lab↗ Design a collage of roadmap metaphors
    32   Science Interview → prof.ir. Ena Voûte

40   **CREATIVE TREND RESEARCH**
    42   Visual Trend Spotting & Niya Stoimenova, Dirk Snelders
    48   Strategic Trend Scanning & Dirk Snelders
    52   Four Techniques for Creative Trend Research & Niya Stoimenova and Dirk Snelders
    60   PEERBY Case & Anna Noyons
    68   Lab↗ Try out the Trend Topics technique
    70   Science Interview → prof.dr. Susan Reid

76   **FUTURE VISIONING**
    78   Future Vision
    84   Vision Concept & Ricardo Mejia Sarmiento
    88   Creative Lead
    94   FERRARI Case & Flavio Manzoni, Enrico Leonardo Fagone
    100   Lab↗ Expression on a desirable future
    102   Science Interview →prof.dr. Gloria Barczak

CONTENTS

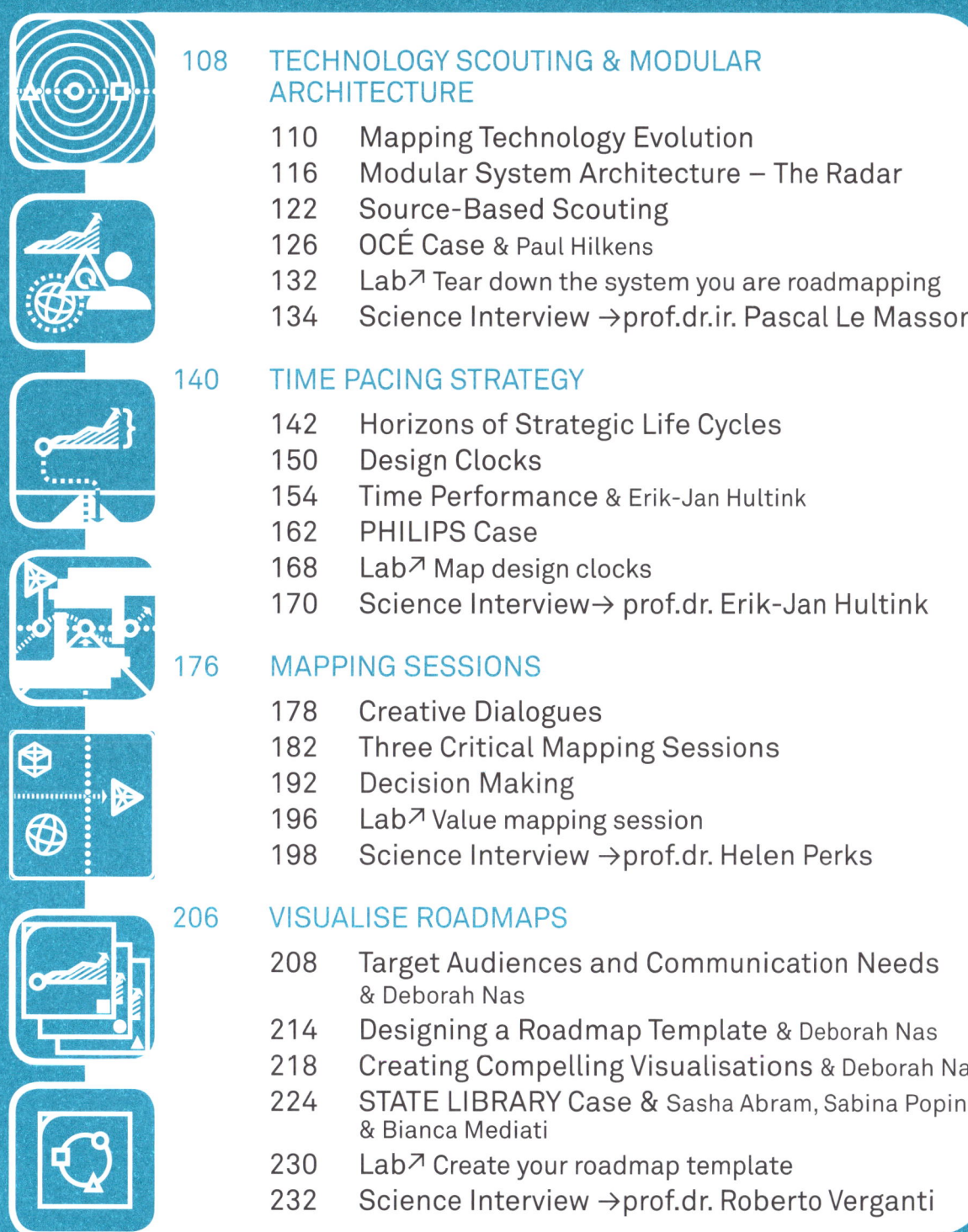

108 **TECHNOLOGY SCOUTING & MODULAR ARCHITECTURE**
- 110 Mapping Technology Evolution
- 116 Modular System Architecture – The Radar
- 122 Source-Based Scouting
- 126 OCÉ Case & Paul Hilkens
- 132 Lab↗ Tear down the system you are roadmapping
- 134 Science Interview →prof.dr.ir. Pascal Le Masson

140 **TIME PACING STRATEGY**
- 142 Horizons of Strategic Life Cycles
- 150 Design Clocks
- 154 Time Performance & Erik-Jan Hultink
- 162 PHILIPS Case
- 168 Lab↗ Map design clocks
- 170 Science Interview→ prof.dr. Erik-Jan Hultink

176 **MAPPING SESSIONS**
- 178 Creative Dialogues
- 182 Three Critical Mapping Sessions
- 192 Decision Making
- 196 Lab↗ Value mapping session
- 198 Science Interview →prof.dr. Helen Perks

206 **VISUALISE ROADMAPS**
- 208 Target Audiences and Communication Needs & Deborah Nas
- 214 Designing a Roadmap Template & Deborah Nas
- 218 Creating Compelling Visualisations & Deborah Nas
- 224 STATE LIBRARY Case & Sasha Abram, Sabina Popin & Bianca Mediati
- 230 Lab↗ Create your roadmap template
- 232 Science Interview →prof.dr. Roberto Verganti

DESIGN ROADMAPPING

9

DESIGN ROADMAPPING

WHAT IS A DESIGN ROADMAP?

A roadmap on the strategy of design innovations takes the future vision as its destination. On the map are pathways that are dotted with new products and services landmarks. For the journey into the future time zone, a roadmap includes parallel tracks – next to the track for design innovation in the existing business, a roadmap introduces the next frontiers of innovation with new value propositions for new user groups. To learn more about the values of new users and prepare for the disruption that a new value proposition may create, the alternative pathways are mapped on the roadmap timeline - There are pathways that enhance the existing business propositions and there are alternative routes towards the exploration of new market spaces and new technology areas. The future timeline provides the backbone of the roadmap. Basically, a design roadmap is a map used to visually track and strategically explore future design innovations.

In this first chapter we will define what a roadmap is, and describe the overall process of roadmapping. We will also take a deep-dive into the history of roadmapping and grasp some inspiration from cartographic roadmaps. You will see that even the earliest roadmaps had extraordinary visual power.

Design Roadmap

↘ Definition of what a roadmap is

Let's start with the definition of a design roadmap. A roadmap is defined as: *a visual portray of design innovation elements plotted on a timeline*[1]. Elements such as user values, new products and services but also market segments, technoloy applications and touchpoints. Each roadmap has its particular format and visualisation[2,7,8]. An example in figure 1.1 shows a service roadmap for a digital food service of a grocery organisation. Design studios all over the world, from Denmark to Australia and from Korea to the US, are experimenting with roadmapping. Each finding their own style in plotting the basic elements of value, product-, market- and technology choices on the timeline [3,4,5]. Figure 1.2 provides a schematic overview of the typical elements included in a design roadmap. Depending on the context and the designer's signature, each roadmap design has its particular format, specific additions and signature visualisation. Figure 1.4 shows another example of a design roadmap.

Roadmaps not only provide strong visualisation and decision making support – they foremost enable organisations and designers to devise creative responses to future strategic challenges[3]. A roadmap supports the innovation strategy of an organisation, because the decision making for a roadmap involves the creation, exploration and convergence of ideas about the future[6]. In essence, a roadmap offers a

A ROADMAP is a visual portray of design innovation elements plotted on a timeline.

Figure 1.1
Digital Food Service Roadmap

cc Esmee Mankers, 2017.
Master of Science Strategic Product Design, Graduaction report. Faculty Industrial Design Engineering, Delft University of Technology.

tactical plan on design innovations to turn a future vision into a reality. The decision making involves innovation professionals at every level of the organisation[1,6,7]. Because articulating the design innovation plans implies uncovering new trends, scouting for new technologies and mapping user values[2]. Together you decide on what future vision suits your organisation. After which more decisions follow on the direction of the innovations, the framing of the time pacing and the concepts needed to reach that vision[1].

Noteworthy is, that it is actually quite rare to see a design roadmap in public[6]. Often, it is a confidential document because a roadmap contains sensitive information that might be of particular interest to competitors, journalists and individual users[7]. Therefore, all the roadmaps in this book have either passed their actualisation dates or they were newly created by strategic design master's degree students as part of their learning process.

Roadmap Design

Just as there is no 'standard' design for cartographic roadmaps – "No two cities in the world draw their maps to the same scale and use the same map legend"- no standard design exists for roadmaps [7,8,9]. As a general guideline, a typical roadmap will take a future timeline and four layers related to the innovation dimensions of: user value, markets, product -service and technology as illustrated in figure 1.2.

The minimal critical specification of a design roadmap is the visualisation of only the product-service layer, with the layers of user value, new market segments and application of technologies explained in words - as exemplified by the Wake-up light roadmap shown in figure 5.13. The maximal specification of a roadmap would include the standard four layers of the design roadmap, and a parallel program layer that translates the design roadmap elements into project activities and its resourcing in manpower and financial investment. As shown in the example of the Quby roadmap in figure 7.5.

A roadmap design of the actionable innovation strategy has three basic characteristics: it is (A) a visual portrait of the organisation's future innovations, (B) outlined by user value, product - service, market and technology elements (C) plotted on a timeline[1].

(A) The 'visual portrait' characteristic of the map concerns the graphic design and its visualisations on the map. Beyond a written strategy document, a roadmap is an eye-catching, informative visual portrait of the design innovation plans. It is a map, that enables stakeholders to more concretely imagine and envisage design innovations. A living visual 'document' - a map that graphically portrays and visualises the future plans of innovation.

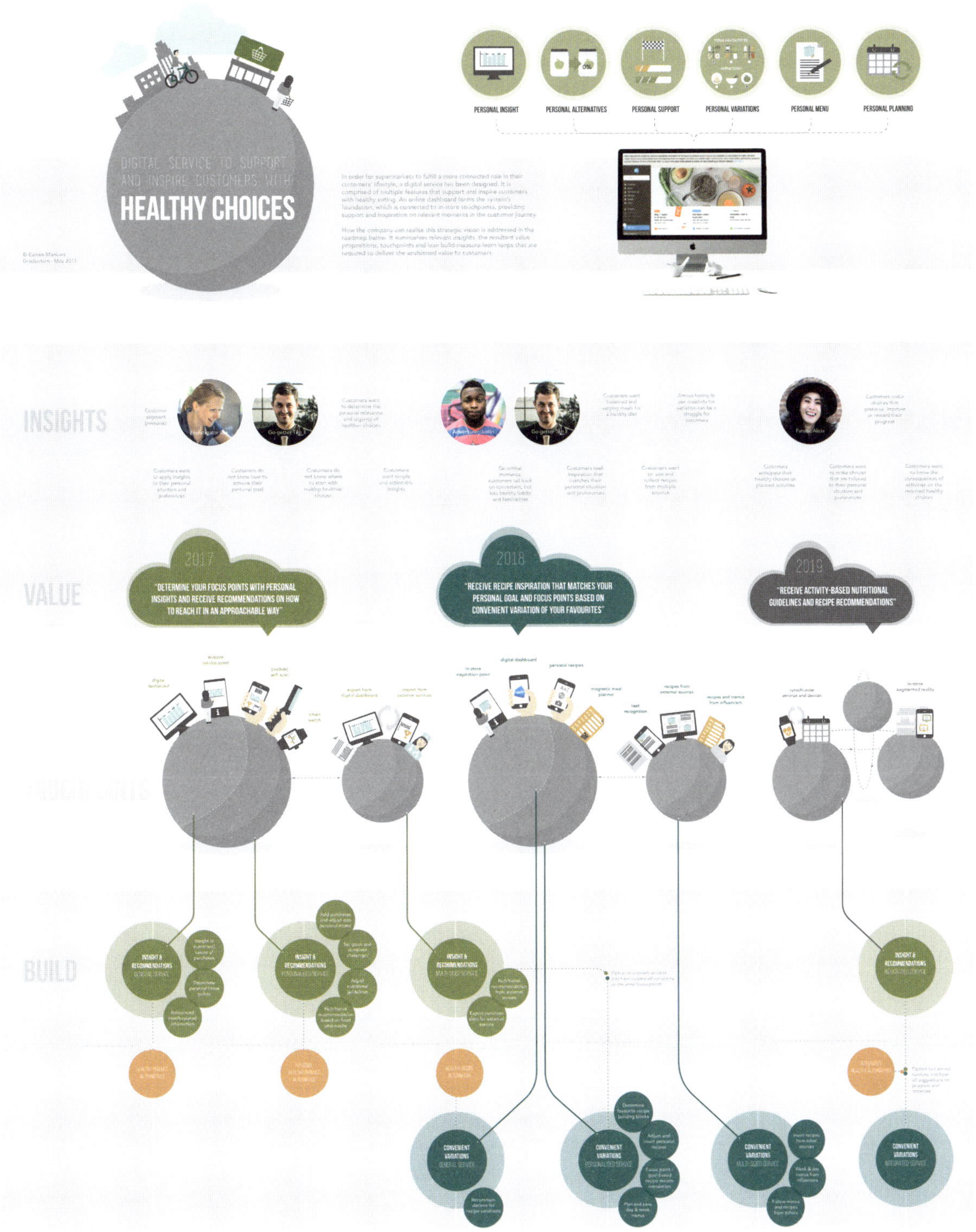

DESIGN ROADMAPPING

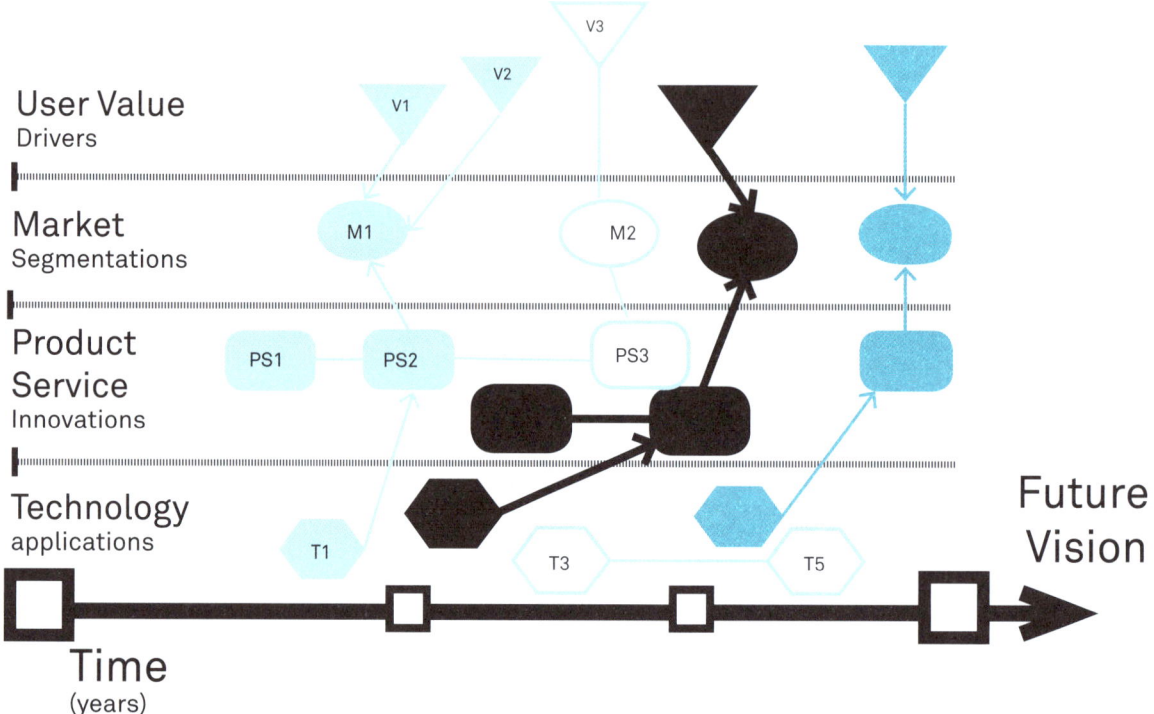

Figure 1.2
Schematic overview of the elements of a design roadmap

cc Simonse, Hultink & Buijs, 2015.

(B) A design roadmap outlines the explorations on the value in the market, product and technology area. A manager at British Petroleum (BP) explained once that a roadmap contains "easy-to-comprehend descriptions of customer needs, technology responses and R&D programs[1]." Roadmaps unravel and connect the different dimensions of the design innovation plans.

(C) The timeline is the most prominent characteristic of a roadmap. "The future timeline embodies the strategy narrative with a beginning, a middle and an ending, and the focal topic of design innovations[1]." The organisation's future vision is plotted at the end of the timeline, while their current business situation acts as the point of departure.

The timeline makes the future time clear (visually) and negotiable for interactions by the various innovation professionals involved. The timeline is the bridge in building and discussing the roadmap, connecting the three (or more) layers of the roadmap. It synchronises the choices and decisions on innovation. Beyond plotting the future progress of incremental product or service releases, a roadmap also anticipates disruptive technology leaps and entirely new markets of users using entirely new platforms, products and services.

Practitioners from the telecom industry were among the first to publish a roadmap example [9, 10]. Figure 4.3 shows a technology roadmap created by Lucent Technologies[10]. This roadmap depicts key market values, product visions and technologies on roadmaps visually organised into 'swimming lanes'. Often, several lines of business including several product-service system lines are mapped parallel to each other. Contemporary roadmaps tend to have a service roadmap at their core [3,4], as illustrated in figure 1.4, and some roadmaps have extra layers that map out social trends, business models and platform system releases.

What a roadmap is not

In trying to grasp exactly what a roadmap is and does, some managers tend to relate them to product portfolios. The main difference between a roadmap and a portfolio is that roadmapping encompasses a time perspective and a set of activities that allow a firm to envision, conceive, select and direct the pipelines of new products that will align with the firm's strategy over the long term. While portfolio management emphasises the commercial feasibility of new innovations with its financial estimations related to business cases profitably, a roadmap articulates a tactical innovation plan that translates strategic objectives into a vision and concept ideas of what to innovate[1]. It shows the 'roads'– including potential alternatives – towards achieving that vision.

What a roadmap is also not, is an implementation plan. It does not plan the implementation phase of whatever single new product or service. Nor is a roadmap a planning suited for tracking progress. As the roadmap example in figure 1.4 illustrates, roadmaps are not created to support day-to-day planning, implementation oversight or operational management. On the contrary, a roadmap serves a long term view over several years and aims to support strategic stability. Opposed to a resource- or project planning, a roadmap visualises a business's strategic direction [3,7,8]. The roadmap is discussed and adjusted only a few times a year – perhaps once per quarter, or every six weeks – or when there has been a disruptive event that requires an innovative response by the firm [2,9,10]. In essence, the roadmap shows the firm's long term, strategic outlook over several years, and the timeline mapping signifies incremental steps or more breakthrough jumps toward design innovations.

Overall, a design roadmap provides a visual means of strategic communication and direction of the innovation program, stable enough to enable coordination across communities of innovation practice and flexible enough to adjust[7,8], to different strategic scenarios of design innovation.

Process of Design roadmapping

Underlying design roadmapping is a process of diverging and converging activities structured in three stages. You can carry out three main mapping activities, starting with 'Value mapping', followed up by 'Idea mapping' and ending with 'Pathway mapping'. As outcomes of your roadmapping efforts, stage 1 aims for a shared future vision, stage 2 a design roadmap and stage 3 a design program roadmap. Depending on your role, the business context and the number of people involved in the roadmapping team, the key activities can be organised and subdivided into tasks to be undertaken by a single person or a group. Team members share their findings and knowledge during one (or more) roadmapping sessions per stage. Additional tasks tailored to a particular context or strategy intent can be added. Figure 1.3 shows the baseline of the roadmapping process – you may want to use it as a guide when organising and preparing the creation of a new roadmap.

↘ Value Mapping

Roadmapping starts with value mapping. First you can conduct a creative trend research, a divergent design research activity intended to establish a future outlook on the environment of your organisation. The trends provide potential directions of new opportunities of value creation. After several members of the roadmapping team have completed their design research, the team converges around an activity of sharing and accumulating knowledge about the future. This team activity is focussed on creating a vision. The mapping challenge is to create a value map that fosters a collective understanding of the current strategic position and to grasp new value opportunities from trends and technologies into a future vision. With the values mapped out, team members can more easily decide which unique value drivers they want to underpin their organisation's future vision. At the end of this stage, the roadmap will have its destination: a future vision statement grounded in unique value drivers.

↘ Idea Mapping

At the heart of the roadmapping process is idea mapping. During this stage, the divergent activity consists of 'technology scouting' at the detailed level of service systems modules, in parallel to more in-depth research into user value drivers. The results of this research provide input for the team activity of idea generation. The mapping challenge here is to generate ideas for user values with matching technology applications. On the draft roadmap prepared for such idea mapping, the time pacing

Stages in the ROADMAPPING PROCESS are value mapping, idea mapping and pathway mapping.

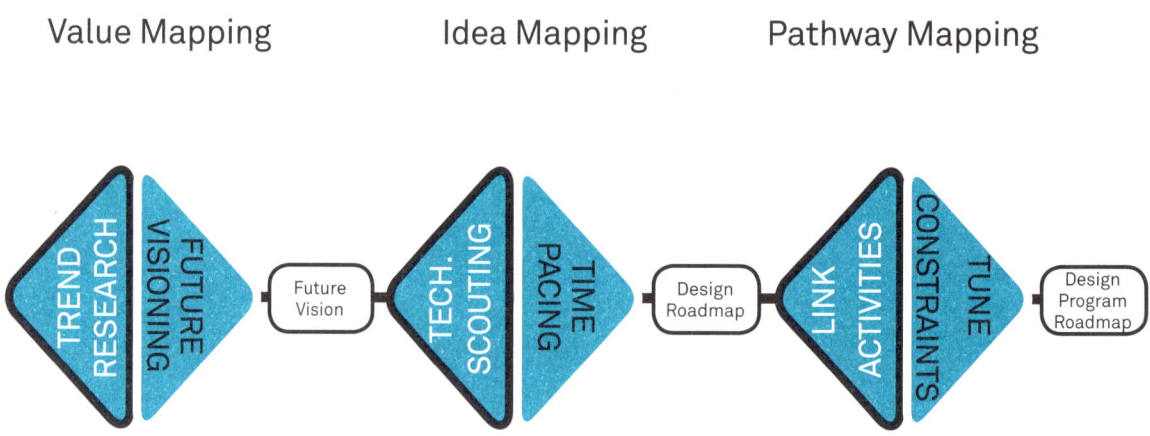

strategy serves as a backbone. The decisions to be made, concern the timing per type of design innovation in connection to the generated concept ideas, user values and technology options. The result of the second stage is the delivery of a design roadmap.

Pathway Mapping

To finalize the roadmapping process, the key activity is the mapping of pathways of innovation activities. During this third stage the mapping will enable the team to flesh out the design roadmap in greater detail, and fine-tune it into a design program roadmap. The diverging concerns the detailing that includes the estimations of lead times, resourcing in manpower and financial investment. For the ensuing convergent activity it is important to know the constraints of the resource investments upfront. Then the team can connect the envisioned product-service launches with the leadtimes for the development activities. They can ascribe lead-times for the development that take proofs of technology into account, and the time it might take to design an application for that technology. The mapping challenge is to overcome the tensions between

Figure 1.3
Process of design roadmapping

cc Simonse, 2017.

the 'market pull' - and the 'technology push' activities by synchronising the innovation efforts along the design activities. With pathway mapping you create different flows of activities for different types of design innovations projects. During this stage, the important decisions for the roadmapping team relate to balancing the amount and type of pathways within the constraints and the mapping of alternative pathway scenarios. The team needs to harmonise the quantity and quality of their innovations and the relationships between innovation pathways, and ensure that all these factors adhere to existing and future resource constraints. The deliverable of the third stage is a design program roadmap.

↘ Roadmapping team

To carry out the roadmapping process you will need more than one person. Typically a team of innovation professionals with diverse backgrounds and roles perform the roadmapping.

When you compose the roadmapping team, invite those who can become active in one of the roadmapping activities including designers, market intelligence experts and technology research experts who can do the creative trends research or technology scouting. Involve product managers, project programmers or people with similar roles who can act as the owners of the design roadmap and design program roadmap. Also involve one or two senior managers – preferably those who decide on the long term development plans to align with their innovation priorities.

We also recommend that you invite persons who have the talent to persuade senior management (vision champions) and those who have the communication and visualisation skills to convince the critical mass of managers on design and innovation. Whenever appropriate think about selecting people for your roadmapping team, who are your organisation's business-to-business clients, strategic partners, preferred suppliers and do not forget to include one or two inspirational 'rebels', who are passionate about future imagination.

Compose the roadmapping team in such a way that you ensure a solid base for creating common ground on the future vision and the design roadmap.

Roadmapping performance

On the subject of what you can achieve with roadmapping, several experienced roadmappers have shared their learnings by documenting roadmapping examples from their organisation and industry. Some of them saw a large impact on the effectiveness of R&D through an increasing customer awareness and successful new products through market differentiation: "We have placed competition ready products in attractive market segments and better determined our technological positioning - Siemens". Others saw an impact on cost efficiency through a reduction in manufacturing costs and accomplishing cost leadership through specific targeted innovations in their highly competitive markets. Roadmapping's "KSFs (key success factors) are time-dependent - they represent hard and specific targets to be attained during the implementation of the strategy and measure of the impact and effectiveness of the R & D – BP[1]."

Almost all of them saw an impact on achieving a better competitive edge by better time-to-market. – a better competitive timing[1]. Furthermore, beyond a single company, in a roadmapping context of an industry network, roadmappers experienced that "combining the resources across companies may make developing the technology possible and consequently the industry more competitive - Sandia[1]." Several roadmappers reported about a roadmapping process among alliance members: "we saw a long-term positive development of the entire branch-Aircraft Devices[1]."

Our research delved into twelve professional roadmapping cases and concluded that there are commonly two performance aspects where roadmapping had an impact on: (a) competitive timing, and (b) industry synergy[1]. Roadmaps can be effective artifacts for your own organisation or for your industry sector.

In a corporate context, roadmapping can boost innovation performance in the marketplace, offering you the chance to improve the competitive timing of your user value innovations. Properly timing the entry of the new value innovations is crucial, and depends on taking the initiative and setting the bar, or being responsive and reacting flexibly. For an industry impact, firms that are actively involved in the roadmapping process can all achieve greater innovation synergy and thus affect the industry's innovation performance across the board.

↘ Competitive timing

Competitive timing is competition-dependent timing of new product introductions in response to the innovation cycles and launch rhythms of rival market players[1]. Competitive timing is not about performance over time in absolute terms; rather, it is relative – it involves the launch timing of new value innovations by the competition.

Generally, strategic managers prefer long response times, because the longer the elapsed time between entry of the first mover and that of later entrants, the more opportunities become available to the first mover to achieve cost and differentiation advantages. Research confirms that when firms research their competitor's activity, they are able to introduce more distinctive products with a higher extend of innovation[11]. Studies on whether first-movers take advantage of these monopolistic benefits

ROADMAPPING PERFORMANCE on improving competitive timing or industry synergy.

DESIGN ROADMAPPING

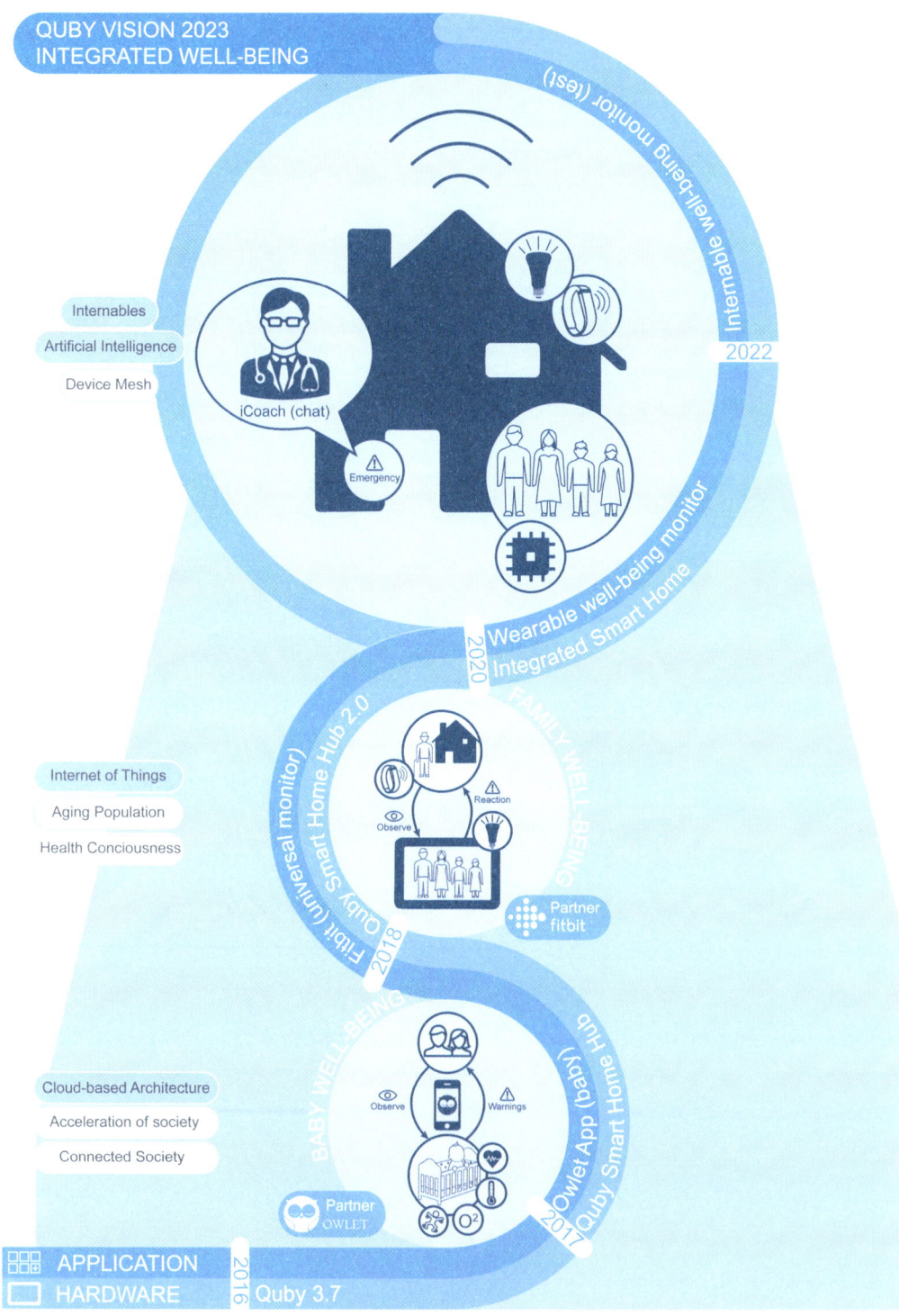

found that success is largely determined by competitors' responses: with a fast response, the benefits are more temporary, while benefits are more durable if there is a slow response[12]. This evidences that competitive timing has a high impact on an organisation's performance.

On average, early and fast movers achieve greater gains (higher extraordinary returns) than late and slow movers. First mover returns also suffer directly at the moment that competitors released imitations of their new products. Sometimes, leading firms also rely on competitive timing when they intentionally wait until a competitor emerges in order to avoid cannibalization of their current products.

Overall, successful firms have been found to have a higher sales volume when they have the ability to get the market entry timing for their new innovations 'right' – neither too early nor too late. These firms benefit from the positive impact of competitive timing with higher returns[12].

Industry synergy

Industry synergy is the value that is created and captured over time by the sum of firms together relative to what they would be separately[1]. Synergy impact stems from the increasing reality of complex product service systems and limited resource availability for innovations. "A certain technology may be too expensive for a single company to support or take too long to develop, given the resources that can be justified . . . it is impossible to independently develop all of the required technologies, technology partnerships can provide a way to leverage these limited resources - Sandia[1]." An industry roadmap allows industry partners to co-develop system technologies, rather than redundantly investing in the same technologies and under-investing or missing other important technologies[1]. One famous example of an industry roadmap is the semiconductor roadmap of ITRS (International Technology Roadmap for Semiconductors) whose alliance partners include Intel, ASML Sandia, Samsung. In the PC market segment, research shows that when Intel took the lead, they influenced the timing of industry changes in such a way that other players – including customers, competitors, suppliers and alliances – joined in to adhere to Intel's time-paced strategy[13]. This exemplifies industry synergy in innovation.

There have been similar roadmapping initiatives in the aircraft and enterprise software industries. Branch networks bring together organisations from across an entire value chain. This kind of 'vertical integration' can offer synergy impact to each strategic partner. From an economic viewpoint, industry synergy represents cost savings or revenue enhancements, but from an innovation viewpoint, it also includes value adding and competitive advantages[1].

Overall, using roadmaps at the industry level relates to better synergy in innovation performance among industry partners.

←
Figure 1.4
Strategic roadmap designed for QUBY

cc Pepijn van Dalen, Luuk Roos & Zoë Dankfort, 2016. QUBY Project report, Design Roadmapping Course, Faculty Industrial Design Engineering, Delft University of Technology.

Please note that the design roadmap is created for Quby by Strategic Product Design Master students, and therefore do not reflect Quby's actual strategy.

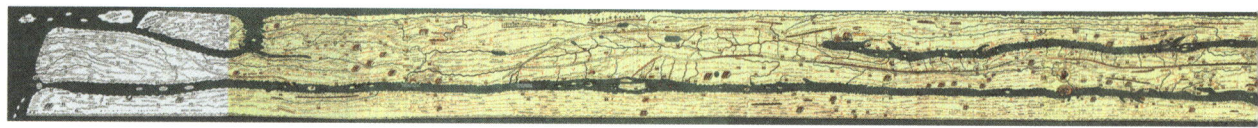

↑↖
Figure 1.5
Peutinger Map,
Dated 63-12 BChr.
Size: 680 x 34 cm.

The roadmap uses symbols to represent landmarks, points of interest and resources; schematic lines with 5 km = 25 km to depict itineraries and routes that led to the metropolis of Rome.

EV   Yeah, kind of. What I want to do is map these cards with respect to a longer-term personnel strategy that I would like to have for the faculty. At the moment, this happens informal and very spontaneous. I think it is not good enough. So I am going to organise a session to have a discussion with the MT first. This way I have visualised it by creating cards with pictures of the people from our faculty staff and a map with roles that need to be fulfilled. The use of pictures makes it instantly clear which discussion is on the table.

> LS   Also because by mapping the discussion, the topic becomes more constructive, and, in a way, you can use the map to structure the discussion?

EV   Yes – and because you *see* a certain structure. Everyone *sees* the structure at the same time, which makes it easier for us all to start from the same position and think more freely about our contribution to the discussion, and enables us to oversee the total plan and possible variants of this plan. And because it is flexible we also do not get stuck on only one solution. With an initial vision on designers of the future as a starting point, groups are now working on a renewed bachelor that prepares our students for finding their way within the complexity of designs of the future. Each new group that starts working on it enlarges the momentum and refines the direction.

> LS   So that is also how you formulate a strategy: first with a small team, then create momentum with a critical mass and then implementation?

EV   Yeah that is true, but before you actually start in a small team I have a lot of input already, to get a feeling about where the space is. In terms of content, what is the 'space' available for the creation of new directions and strategic constructions, but also 'space' in terms of the meeting of emotional minds - what is the emotional space that people will give each other, what is their enthusiasm and energy drive and where do I find traction.
    The choices that I need to make are different from industry situations in which you sometimes have no choice – you have to tackle a very difficult problem, in the academic world we are in the fortunate position of being able to contribute in (too) many places, allowing us to make strategic choices based upon our own interests and capabilities. I still like to have a few focus areas where we can put most of our energy. Therefore, strategic focus areas are in place, so our scholars can join forces and be seen and regarded for their work. Clustering our energy

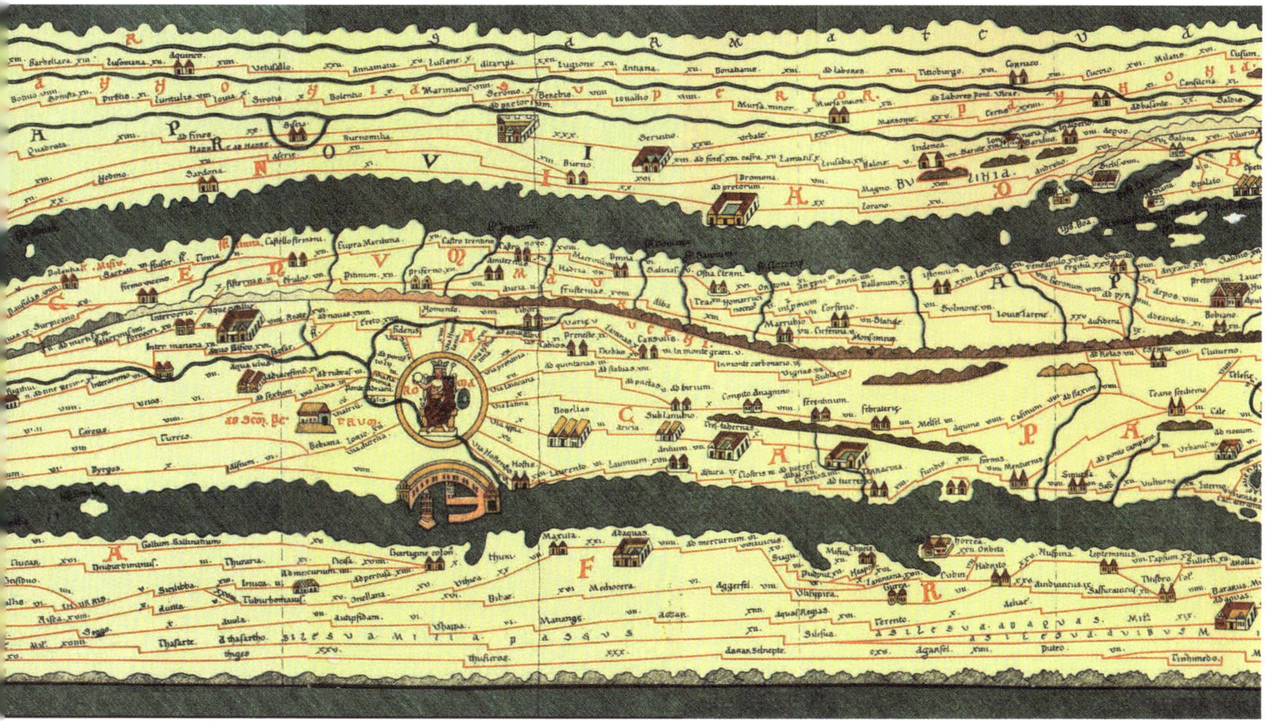

25 DESIGN ROADMAPPING

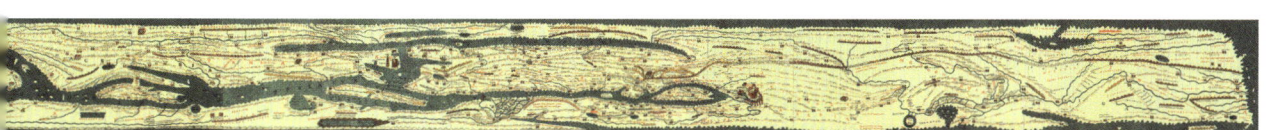

METAPHORS CASE

Roadmap metaphors

In Merriam-Webster's online dictionary, a roadmap is defined as
1. a map showing roads especially for automobile travel
2a. a detailed plan to guide progress toward a goal
2b. a detailed explanation[14.]

The primary meaning of the word 'roadmap' – a map for navigating in a car, taking us from A to B and showing alternative routes – provides a strong metaphor for a design roadmap: in which strategy provides a journey, visions are the destinations, means are routes, and service achievements are landmarks. These kinds of figures of speech have inspired numerous roadmappers in relating cartography to innovation strategy. Metaphors have the capacity to open up new ways of understanding[15.] In this section we share five cases of historical roadmaps that did far more than suggest which route was possible. First we go back to the oldest roadmap in history, then introduce a circular map of the world, examine a colourful example of a world famous cartographer and, finally, take a leap forward to today's digital roadmaps, which enable us to navigate and visualise our future journeys in real time.

↘ ### Peutinger Map

On our deep-dive into the history of roadmapping we found the Tabula Peutingeriana or Peutinger Map (figure 1.5) the oldest roadmap of European heritage named after Konrad Peutinger, dating from about 50 years before Christ. The Peutinger Map is a parchment scroll, it stretches from west to east over 6 meters and 80 cm by 34 cm.

This map with every road literally leading to Rome, has an extraordinary visual appeal. It was originally designed for the pilgrims of France to guide their travels along routes featuring visual icons that represent the orientation points and landmarks of cathedrals, towers, rivers, forests, mountains, thermal sources.

Similar to a contemporary subway map, the lines for each route are drawn clearly, yet paid no heed to mathematical scale or geographic precision, the essential being their indication of distances and important crossroads rather than topography. All of this makes the Peutinger map a visually inspiring metaphor for design roadmapping.

↘ ### Mappa Mundi

Another famous roadmap is the Mappa Mundi, which dates from around 1300 CE[16]. The circular map has Jeruzalem as orientation point in the centre (see figure 1.9). The world that extends outward is visualised as round and flat. Besides mapping the geographical continents, countries and cities including Europe, Asia and North Africa, at its edges – the map also features people, animals and creatures of the after world, its depictions of our destiny resembles a kind of imagining of the time after the map's creation – or the time beyond the map. In a metaphoric way, it can be associated to a design roadmap, which is itself a depiction of future imaginings. Although, a timeline element that is typical for design roadmaps, is not explicitly part of the mappa mundi. The map covers the biblical story time, from creation to doomsday.

At Hereford Cathedral in the UK, the mappa mundi is on display, measuring 1.59 by 1.34 meters. East, where the sun rises, is at the top. Countries and oceans are squeezed and stretched to fit into the map's circle. The mappa mundi reflects the thinking of the medieval church. The inhabited part is shown– here Caesar Augustus, there farmers harvesting corn and over there performers dancing with bears. But then, the map also contains images of Adam and Eve, Noah and his beasts and a man riding a crocodile. Short descriptions offer small nuggets of wisdom like "Here are strong and fierce camels." Some of this information came from travellers and written accounts, and some apparently came from pure imagination, such as those funny creatures with huge ears wrapped around themselves. More than a reference for geography, the Mappa Mundi is a work of the imaginative world.

↘ From earthbound cartography to a bird's eye view

The roadmapping art work of Dutch cartographer Joan Blaeu is another noteworthy source of inspiration. Blaeu impressed many audiences with his quality and art work. He was also known as an innovator, as he was one of the first to use the newly invented printing process to produce high quality colour maps. The map of Delft is part of the Atlas Maior, which measure 37,5 by 49 cm.

Out of curiosity, we decided to compare Blaeu's excellent mapping to today's cartography techniques with satellite and aerial mapping images. Nowadays, a satellite or aerial image provides spatial resolution ranging from 15 meters to 15 centimetres. In figure 1.6 a Google Earth image of the Dutch city of Delft is shown below Blaeu's hand drawn map from 1649. The accuracy of Blaeu's work seems to have stood the test of time! Despite being an important factual resource, it is also a work of art. It is a source of information and inspiration. Google's aerial image is arguably more accurate, and still despite the sharpness of its lines and the height of its resolution, it leaves room for us to imagine how people live inside the houses and buildings we see.

The Map of the Internet

A more schematic map that is associated to the digital revolution is the map of the inception of the Internet. Figure 1.7 shows the schematic representation of the Internet. It connects the Advanced Research Projects Agency Network (ARPANET) in Virginia with the network of Stanford Univeristy in the west the US, to the MIT and Harvard networks on the east further across the seas to London and in the west to Hawaii.

The schematic overlay's squares and ellipses resemble a sort of basic type of subway map that has no intention of representing the exact distances the data packets would be travelling – it merely existed to explain how the networks would interlink. Visible interlinkages are also a crucial part of a design roadmap's composition.

The year 1974 of the Internet map demarcates the invention of the TCP/IP by Vinton Cerf and Bob Kahn. By a long run of 15 years later Tim Berners-Lee, then scientists at CERN, designed the World Wide Web concept in 1989. According to Berners-Lee "The Web is an abstract (imaginary) space of information. On the Net, you find computers – on the Web, you find document, sounds, videos and information.... The Web made the Net useful because people are really interested in information (not to mention knowledge and wisdom![17]". It is interesting to note that the two perspectives provided by 'the Web' and 'the Net' are necessarily intertwined, and interlinked. In a way, roadmaps do the same, interlinking multiple perspectives.

Infographics

We are living in a new area of design, when our creative and design abilities can be augmented with the power of big data. Although there are many more metaphors we could explore, lastly we introduce an infographic map by Francesco Franchi. He designed an impressive visual representation of global high-speed train activity. Big data visualisations often begin with a geographic silhouette – this makes the (localised) data more understandable and accessible, by telling a complex tale in a simple, visual manner. In the example shown in figure 1.8, Franchi uses a map to support his visualisation of the routes high-speed trains take in different countries. The colour scheme visualises presently operational and non-operational routes, in addition to planned future routes. On a design roadmap we also have several layers – we visualise the current business, the growing business for which the development has started and the future business on which research is taken place. The circular layout that Franchi uses in his high-speed train infographic could be an inspirational source for a layout for a design roadmap.

We have presented all these metaphors to encourage you to create your own roadmap design.

Figure 1.6
Joan Blaeu's Roadmap of Delft.
Dated around 1649,
Size: 37,5 by 49 cm

Google Earth's image of the city center of Delft, the Netherlands.
Dated: 2017.
Size: adjustable.

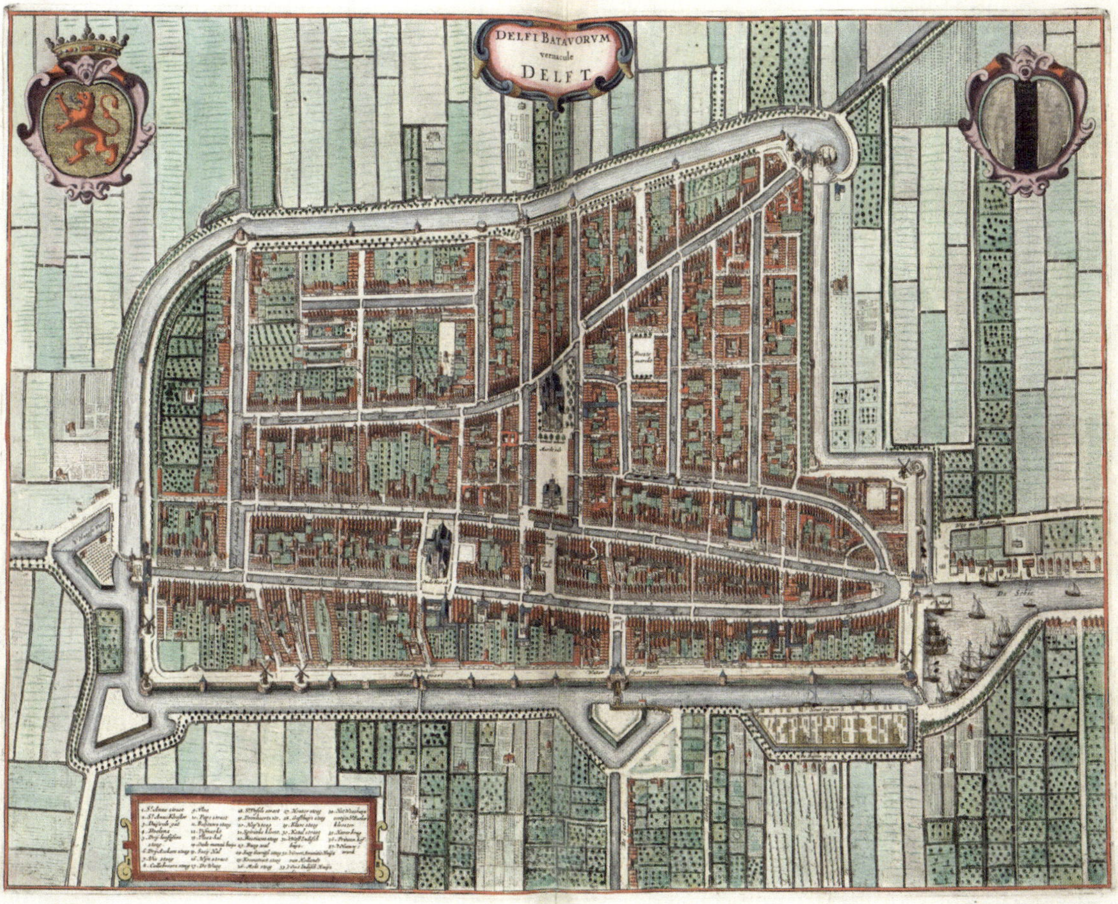

LAB ↗

Design a collage of roadmap metaphors

MATERIALS NEEDED:
- → access to the Internet to conduct a web image search
- → roadmap image matrix : 6x6 rows x columns digitally created.
- → blanco sheet for the collage - for instance an One Note sheet.

1 Collect roadmap images and metaphors from the web that appeal to you. Use the search engine image view to pick visuals you like.

2 Fill out your roadmap image matrix with images that appeal to you for each of the 6 innovation elements on the roadmap: User Value, Market, Product-Service, Technology, Timeline (see figure 1.2 for the roadmap elements). Use one row per element and make a collection of several images per element until you completely filled the 6x6 matrix.

3 Pick 6 favourites out of all the images in your roadmap image matrix. Which 6 elements really stand out to you in your image matrix?

4 Take the sheet to portray the centerpiece of your collage. Pick one of your favourite elements, and give it centre stage on your sheet. It's up to you how much space on the paper you use, but remember that this will be the primary focus of the collage.

5 Arrange the other 5 favourites from your matrix and portray them around the centrepiece on the collage. Things to think about: do you want to arrange it symmetrically or asymmetrically? Would you draw two identical versions or two variations on the same theme? Are the images similar to your centrepiece or do they create contrast? There's lots of room to play here, so have fun!

6 Choose another set of images from your matrix, and arrange them anywhere on the collage. You can take the same approach as you did in the last step. At this point, your composition should be coming into its development.

7 Create or draw a pattern that connects or unifies the images on your collage, touching elements from each area. Think of this as a way to give an extra layer of background detail to your collage.

8 For the remaining white space, mentally divide the collage in half, then finish each half by colouring, rearranging, removing and adding images. Keep it simple. This final step will help you make your composition saturate the collage in subtle ways.

Creating a collage that centres on metaphors will encourage your ability to distil complex ideas into digestible sets of images. Starting with a simple, hands-on approach using the image search in your web browser, here is a perfect opportunity to prepare for your design roadmapping skills.

This is the kind of creative activity that you could easily do in one sitting, probably in less than four hours. The objective of the Lab is to activate your design roadmapping ideas and generate your first visual ideas for design roadmaps.

Prof. ir. ENA VOÛTE on the subject of Strategy and Design Roadmapping

LS Our faculty strategy is called 'Design for Our Future'. You chose to have *Future Foresight* as one of its cornerstones. Why?

EV I think as a designer you can be extremely productive when you express a vision and put it on a map. The map makes what you are thinking about clearer, generates enthusiasm or feedback, and gets stakeholders on board. From the abstract, vague and rather unknown clues on what turn the future might take, we have the skills to creatively express a future vision and make it concrete. We visualise it, so everybody gets a sense and feel for the ideas and directions that we want to pursue. Foresight combines our creativity, our integral approach and our values with our practical skills of engineering. It is an excellent role for a designer to play. To me, future foresight is not about designing a dream – it's about designing a dream that has 'hands and feet', as we say it in Dutch. – That means that we make it evident, that people can see that *it can work*. Future foresight is, therefore, not just an ideal – a kind of utopian view of the future – on the contrary, I think that a very idealistic story is not a design as we see it. In future foresight we build views that we believe can be achieved over time. They can still be conceptual but concrete enough to be able to imagine it can happen.

LS Future foresight itself is also a new area in the field of design. What makes it distinctive?

EV Future foresight is part of our strategy, first of all because in the areas of contemporary society where technology plays a large role in changing our way of living together. With future foresight, we see the problems where we can help, and because we understand what technology can do, we know where we can contribute with solutions. We can build our foresight upon educated guesses, scenarios on which turns the future may take. Because we are very often ahead of the pack and not only do we understand the possibilities that technology offers, we have learned how to figure out how much space people give – or do not give – in their willingness to accept new innovations. We know how technology will connect to people when we design future foresights.

    Second, future foresight relates to a unique position for design. We integrate 'software and hardware,' people and technology. We think integrally, and add future aspects to other disciplines. Psychologists are very good at pinpointing problems, but sometimes they love help to create solutions that actually solve the problem. Economists know how to earn money, but not necessarily with what. Technology experts sometimes forget people. And philosophers could benefit from our visual abilities in communication. So, as designers, we can add by making future visions concrete in such a way that people can actually follow the analysis and forward thinking. In strategy, one always has to bring a lot of different people together to get things moving. The integrative role of future foresight is a role I think we designers can apply for.

> LS In roadmapping, besides formulating a future vision, we also investigate the organisation's past performance in terms of business activities. How do you think about taking inspiration for the future, if you compare that to investigating the past?

EV There are different levels to build future foresights in creating a roadmap. On one hand, you can look at the past, when you are looking for incremental changes in design innovations. Because the past reveals the pace of change and the moments when critical masses created space for new things, or when consumers began to adopt new things. It is also valuable to investigate how people perceive things at the current moment in time. A car for example is perceived as having four wheels, and when it has not and looks very different, consumers might not see it as a car. On the other hand, if you have totally new technologies coming in, like the electric motor, they open up a future of new possibilities that gives us the freedom to design, for instance a completely different look. New technologies often create fresh options for new design geometries and different kinds of functionality, as you can see with the self-driving cars. Then you can actually turn around and can start designing from the future, from a green field and see what the innovative possibilities are. So both investigating the past and exploring totally new options from the future have a value in building up future foresight with a roadmap.

I have also experienced that different types of roadmaps lead to different types of integral innovation. When I was working in industry, I have designed a light armature roadmap for Philips consumer lighting for instance. Based on the consumer insights that people have their own style of creating atmosphere with light, our roadmapping challenge was, how to service many different styles without creating equally many lighting variants, but to service with a handful of items. We designed then new style-architectures per type that could serve various retail channels and reduced the complexity in style choice and certainly in the amount of SKU's (shop keeping units).

Nowadays the possibilities offered by the Internet of Things (IoT) provide a source of future inspiration. In the lighting context every light point is a point in an infrastructure network that connects rooms, buildings and even cities. This network can be a backbone for all sorts of IoT applications.

LS Concerning the roadmapping process, that is organised into sessions during which the roadmap is created together with other innovation professionals. What in your experience, are the key elements to consider when organising a strategic session?

EV Depending on the people you wish to involve, there are different ways to organise a strategic session. You can take the process route, and clearly explain that 'this is the process': we are going to do these activities, we will take those steps, and do this group work to fulfil the process. How you conduct the process depends on whether people are enthusiastic and if they are able to accomplish the steps in the process.

Sometimes it is just good 'to put a point on the horizon', and say 'this is what we can do', and then see what reactions it provokes. Which route to take depends on the aim of the session and the personalities involved. Much relies on the composition of the team, the freedom they sense and the input they can and will bring. What I usually do is try to be very transparent about the aim and the role I would like to take. I love to work with an MT (management team), where open discussions are possible and where there is always room to change outcomes for the better. Ideally this feels like 'designing-on-the go as a team'. Being a design faculty this is a process we are often able to follow and we enjoy it.

LS On your table over there, I see some cards with pictures. Are you going to map them in a session?

is necessary to create enough impact with our work. So, in essence, I organise by building on 'pockets of energy' that move in an orchestrated more focussed manner.

LS   Then my last question, what do you consider as the benefits of roadmapping in a strategy process?

EV   I believe it does a few things. The innovation strategy process can be structured in a few steps so people involved know what is coming. The content that flows through the process however is subject to a design process and as such is growing on and with the people involved. This leads to a shared roadmap with buy-in.

ENA VOÛTE is dean of the faculty Industrial Design at Delft University of Technology. She is professor of practice, graduated as Industrial Design Engineer at the TU Delft, after which she worked for Unilever in several European countries, where she developed launching, branding and marketing strategies for products including Magnum, Becel and Lipton. At the start of this century, she was marketing director at the financial services comparison site Independer, worked as an innovation consultant at Altuïtion, and worked for nine years at Philips in strategy and management roles for Consumer Lifestyle, merely on personal care products and for Philips Lighting on end user and market strategies.

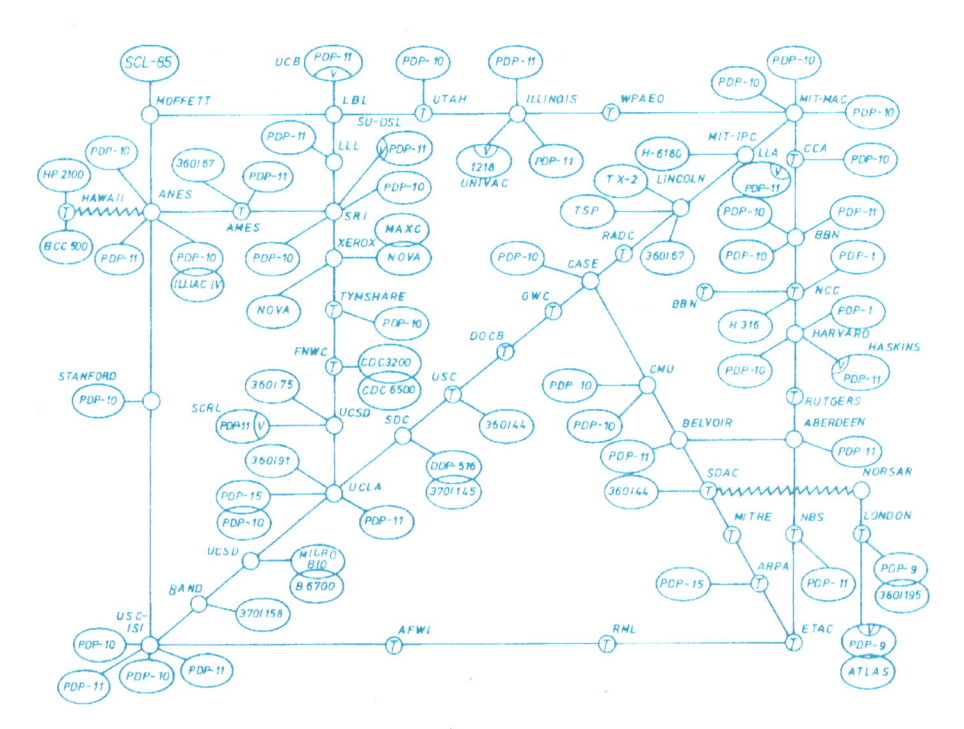

Abb. 4   ARPA NETwork, topologische Karte. Stand Juni 1974.

37 DESIGN ROADMAPPING

↙
Figure 1.7
The Internet Map,
Dated 1974.
cc ARPANET

↓
Figure 1.8
infographic map of global high
speed train activity
Dated 2008.
cc Francesco Franchi.

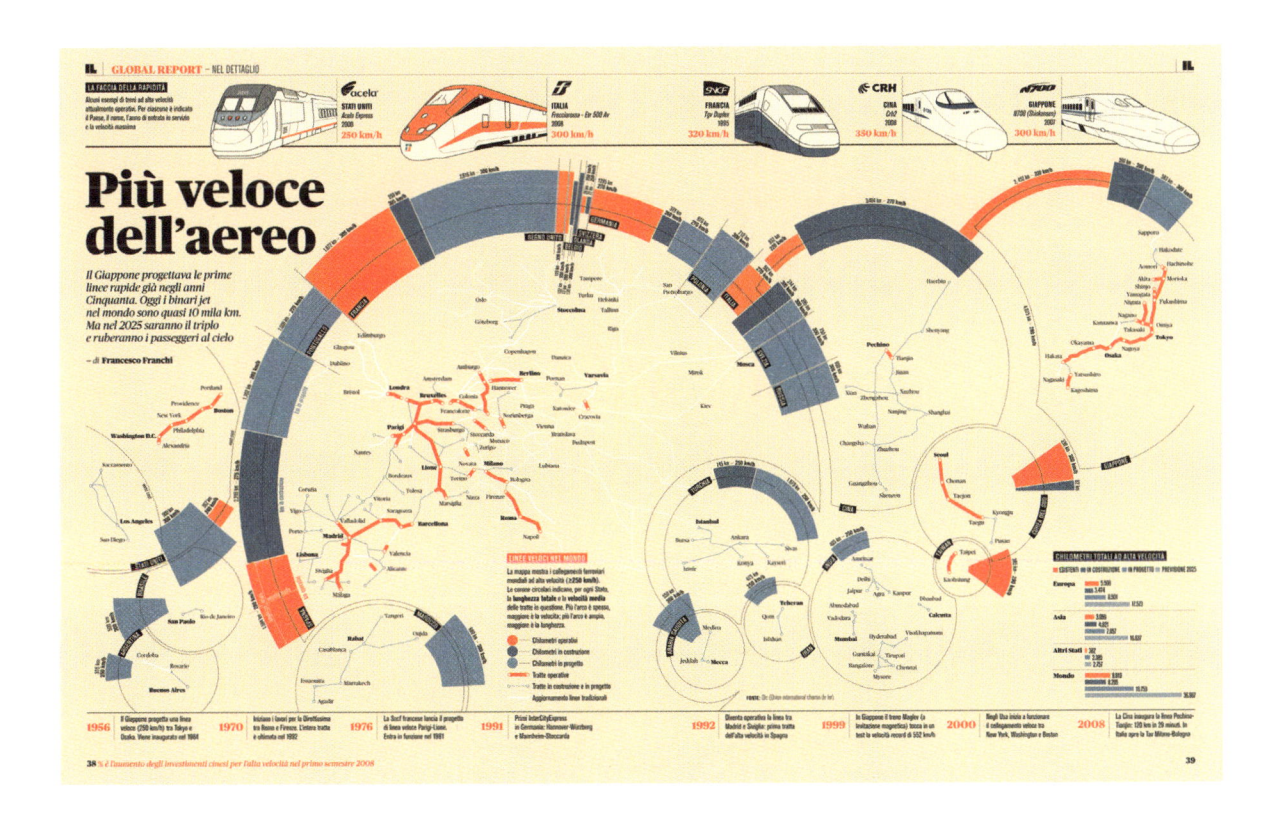

Figure 1.9
Mappa Mundi,
Dated around 1300 CE
Size: 159 x 134 cm.
cc Hereford Cathedral, UK

The roadmap features Jeruzalem as its orientation point in the middle and geographical cities and countries of Europe, Asia and North Africa, at its edges. It also visualises people, animals and creatures of the after world.

IN SUM

We would like to conclude with an answer to the question prompted at the beginning of this chapter about what a design roadmap is.
We began with this definition:

→   A roadmap is a visual portray of design innovation elements plotted on a timeline.

The design of a roadmap is a team effort of multiple innovation professionals. The timeline and pacing are crucial – they synchronise and harmonise innovation decisions across functions, ensuring that innovation stakeholders from every area of the business are on the same page and concentrating on the same goals.

→   The design roadmapping process is organised in three stages, of value mapping, idea mapping and pathway mapping, each of which includes the divergent and convergent activities.

The respective deliverables of design roadmapping are a future vision, a design roadmap and a design program roadmap. Together you carry out the roadmapping process in a team, and create not only an image of the future vision but also the innovation pathways that the organisation can employ to attain that vision.

Ultimately, we encourage you to develop your own signature roadmap in a process of co-creation with your roadmapping team. We offered you the minimal critical specification that a roadmap requires. We showcased a few roadmapping metaphors to trigger your design imagination, and offered a 'Lab' in which you could try yourself to compose an initial draft of a roadmap design. On the importance of the strategic process, our first interviewee (TU Delft's Dean of Industrial Design Engineering) emphasised the inquiry of the social and practical 'space', and the design of maps to structure to it as well, all necessary when you are orchestrating the contributions of an intelligent and enthusiastic group of professionals.

---

1   Simonse, L.W.L., Hultink, E.J. & Buijs, J.A. (2015). Innovation roadmapping: Building concepts from practitioners' insights. Journal of Product Innovation Management, 32(6), 904-924.
2   Phaal, R.C.J.P., Simonse, L.W.L. & Den Ouden, E.P.H. (2008). Next generation roadmapping for innovation planning. International Journal of Technology Intelligence and Planning, 4(2), 135-152.
3   Kim, E. (2016). Design Roadmapping: Integrating design research Into strategic planning for new product development. Doctoral dissertation, Berkeley: University of California.
4   Cho, C. & Lee, S. (2014). Strategic planning using service road-maps.The service industries journal, 34(12), 999-1020.
5   Caetano, M. & Amaral, D.C. (2011). Roadmapping for technology push and partnership: A contribution for open innovation environments. Technovation, 31(7), 320-335.
6   Cooper, R.G., & Edgett, S.J. (2010). Developing a product innovation and technology strategy for your business. Research-Technology Management, 53(3), 33-40.
7   Phaal, R., Farrukh, C.J.P. & Probert, D.R. (2004). Technology roadmapping – A planning framework for evolution and revolution. Technological Forecasting & Social Change, 71(1), 5–26.
8   Groenveld, P. (1997). Roadmapping integrates business and technology. Research-Technology Management, 40(5), 48-55.
9   Willyard, C.H. & McClees, C.W. (1987). Motorola's technology roadmap process. Research Management,30(5),13-19.
10 Albright, R.E. & Kappel, T.A. (2003). Roadmapping In the corporation. Research Technology Management. 46(2), 31-39.
11 Katila, R. & Chen, E. (2008). Effects of search timing on product innovation: The value of not being in sync with rivals. Administrative Science Quarterly, 53(4), 593-625.
12 Langerak F., Hultink, E.J. & Griffin, A. (2008). Exploring mediating and moderating influences on the links among cycle time, proficiency in entry timing and new product profitability. Journal of Product Innovation Management, 25(4), 370-385.
13 Müller-Seitz, G. & Sydow, J. (2012). Maneuvering between networks to lead – A longitudinal case study in the semiconductor industry. Long range planning, 45(2), 105-135.
14 https://www.merriam-webster.com/dictionary/roadmap, accessed July, 2017.
15 Cornelissen, J.P. (2005). Beyond compare: Metaphor in organization theory, The Academy of Management Review 30(4), 751-764.
16 http://www.themappamundi.co.uk/explore.php.
17 https://www.w3.org/People/Berners-Lee/FAQ.html - accessed July, 2017.

CREATIVE TREND RESEARCH

41 CREATIVE TREND RESEARCH

HOW TO DO CREATIVE TREND RESEARCH

With the emergence of future foresight in the discipline of design, the attention for creative trend research has increased[1]. It is also a trending topic online: several leading design bloggers believe that creative trend research and future thinking are now needed more than ever. In roadmapping, we practice creative trend research, which combines the designer's craft of intuitive observations with the strategic scanning of the environment.

The concept of 'trend' has gone through several generations of meaning[2]:

→ FIRST, the term 'trend' is traceable back to the field of geography in the late 16th century, when the term 'trend line' was used to describe the general direction that a stream, coast, or mountain range tends to take. 'To trend' is derived from the Old English trendan and the Germanic trundle, which mean: to revolve, rotate, turn in a specified direction.

→ SECOND, studies of demographic data used the noun 'population trend' in a figurative sense, to determine a trend as a statistically detectable change in collected data on a population. Researchers developed trend analysis tools, which use large collections of quantitative data. For linear trend extrapolation, mathematicians advanced the trend analysis techniques, on which many forecasting studies heavily rely.

→ THIRD, studies on comparing groups of people in the social sciences gave rise to the term 'behaviour trend' that was used to refer to a pattern of collective, socially influenced behaviour. In the 1960s, human-related trend research also became common practice in the field of design. The word 'trend' also became associated with a person: 'trendsetter'. Interior and fashion designers were among the first to be considered trendsetters. Then, the term 'trend-spotter' was coined for people who observe, seek, or predict changing tides in media and material culture (fashion, design, etc.).

→ FOURTH, social media technology ushered a new generation of trend term usage. 'Trend hypes', 'trending topics', and 'top trends' have become part of the daily communication referring to postings on for instance, Twitter, Youtube, and Instagram that are viewed thousands or even millions of times a day.

Considering all these disparate meanings, we concentrate in this chapter on trend research in the creative fields of fashion, architecture, and industrial design. To offer further guidance on how to carry out trend research, we present a framework that includes four techniques: trend scenarios, trend views, trend topics, and trend patterns.

Visual Trend Spotting

LIANNE SIMONSE, NIYA STOIMENOVA & DIRK SNELDERS

Often observed from a few weak signals, which are mostly visual in nature, trends are made by people but uncovered by designers. Design research detects the early signals of a vogue, swing, or drift in trend research:

→ VOGUE: a current style or preference, such as a new fashion trend.
→ SWING: a social movement, such as the shift towards food truck dining.

CREATIVE TREND RESEARCH

VISUAL TREND SPOTTING is counting instances of observed clues over time.

DESIGN ROADMAPPING

→ DRIFT: a prevailing direction or inclination, based on an attitude or preference for one thing over another, such as the drift from living in the countryside to the town relates to the urbanisation trend[2].

Some designers stress that they rely on their intuition in detecting the clues, while others watch for confirmation from multiple sources before they call something a trend. We encourage you to back up your intuition by counting instances of clues you observe over time and then cross-checking your personal observations with other evolving data.

"Detecting the existence of a trend required no more than an intuitive plotting of points on a line . . . showing the direction of the fashion trend."

WILLIAM REYNOLDS[3]

William Reynolds, the renowned theorist on fashion trends, showed us how to make intuition explicit. He plotted observations as points on a line, defining them as evidence of intuitive observations. One of his examples presented the trend of the rise of pop art. By counting the inches of visual print advertisements over time, he generated evidence for his initial intuition about this rise of pop art. Another of his examples concerned the fins of Ford and Chevrolet cars in the 1960s; he plotted trend lines on both the fins lengths and their heights over the years. But his most famous example concerns the trend towards miniskirts. He measured the length of women's skirts on the images in magazines and revealed that skirts became shorter over the years[3].

In daily life, we can intuitively detect hundreds of similar trends. Some are easily quantifiable, based on a strong track record in data from the past. We call these strong signals. Yet others may be subtler in signalling their importance for the future. These so-called weak signals are better examined using sensitive methods that rely on empathic attitudes and the intuitive filtering of observations in the present[4]. Or, as the trend director Hanne Caspersen at Philips Design expressed it, "The world is a jungle of trends. There is no end to them. We don't create them, people make them. Our job is to curate them."[5]

Building on the premise that trends indeed are made by people makes the 'trend setters' the people to watch for. We can seek and spot growing groups of early adopters and opinion leaders, until at a certain moment a critical mass of social influence is reached, and a larger majority of followers comes behind.

←
Americanos -Milano

© Ari Versluis & Ellie Uyttenbroek, 2011: Exactitude 133.

Photographer Ari Versluis and profiler Ellie Uyttenbroek classify random people whom they see in cities around the world according to particular characteristics of their appearances and attitudes. They create categories which comprise of people who share the same attributes, and give each category unique names. This is the artwork they developed together for more than 20 years. They call their series 'Exactitudes ': a contraction of exact and attitude. By registering their subjects in an identical framework, with similar poses and a strictly observed dress code, Versluis and Uyttenbroek provide an almost scientific, anthropological record of people's attempts to distinguish themselves from others by assuming a group identity.

We feel fine

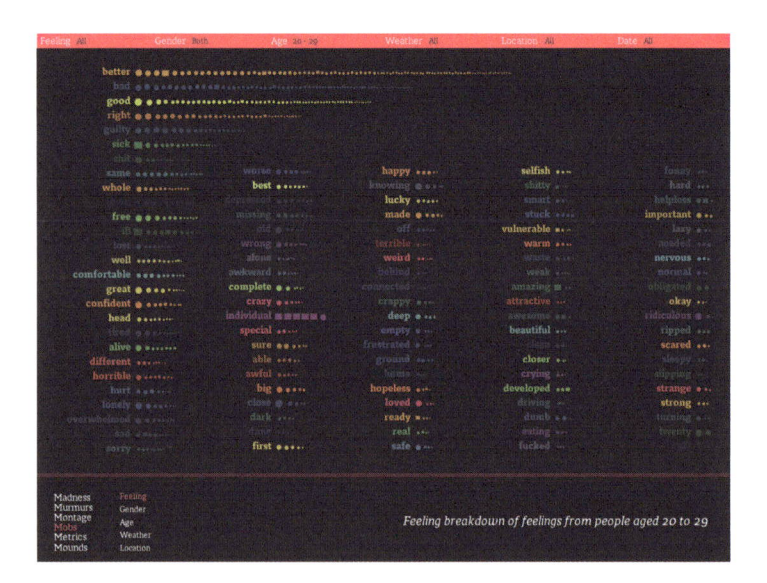

© Jonathan Harris and Sep Kamvar, 2006.

We Feel Fine is an exploration of human emotion on a global scale. Since August 2005, We Feel Fine has been harvesting human feelings from a large number of weblogs. Every few minutes, the system searches the world's newly posted blog entries for occurrences of the phrases "I feel" and "I am feeling". The result is a large database of human feelings, increasing by 15,000 - 20,000 new feelings per day. At its core, *We Feel Fine* is an artwork authored by everyone. It will grow and change as we grow and change, reflecting what's on our blogs, what's in our hearts, what's in our minds.

Typically, the visual trend spotting technique is characterized by use of observations, including the social influence that can be traced back in media and material culture (i.e., conspicuous consumption of fashion, design, etc.). We categorised three types of trend spotting in the fashion and design literature:

→ VISUAL STREET SPOTTING. This technique mainly uses visual research by observations and photos[6.] Some design studios offer this trend spotting as a dedicated service for design research. For instance, Trendzoom and the Doneger Group promote 'Street Trending' as an approach to trend monitoring on the street or other places of social interest[8]. The results are 'a collection of visuals from various global markets around the world'.

→ PROFILING TRENDSETTING CONSUMERS/FASHION OPINION LEADERS. The role of 'trendsetter' has been characterised as 'opinion leader', because the two terms define similar behaviours. Research found that in fashion, opinion leaders consider 'uniqueness' an important factor when choosing what to wear. This is far less important for opinion seekers, for whom, on the other hand, 'information on status consumption' and 'social comparisons' is much more distinctive[7].

→ SOCIAL MEDIA SPOTTING. This technique concerns a special interest in the use of social media technology for trend spotting. Initial findings on the use of blogs in combination with the output of text mining software show promising alternatives to street trend monitoring[8]. New trend methods derive trends from user-generated content from virtual communities such as Myspace[9].

Overall, these three techniques rely on observations. In general, the object of observation is a social movement in fields such as fashion, architecture, and design that can be sensed intuitively. Trends can be observed in daily life, for instance, in newsmagazines, blogs, and speeches. Taste and style trends cohere in the narratives of opinion leaders and through visual material clusters. The detection of trend patterns involves 'fingerspitzengefühl', or a situational awareness, and an intuitive ability to detect emerging trends of the future that can already be felt today.

Strategic Trend Scanning

LIANNE SIMONSE, NIYA STOIMENOVA & DIRK SNELDERS

Strategists introduced 'environment scanning'—the systematic scanning of an organisation's business environment for relevant information[10]. The purpose is to ensure an organisation does not miss out on early signals of possible changes in the environment. Designers use strategic scanning to detect trends of change already underway.

For strategic decision making, designers and futurists crafted the trend technique STEP—an acronym for Social, Technologic, Economic, and Political trends. You can use STEP to scan signals in a global context according to Social and demographic developments, science and Technology inventions, Economical developments, and Political and regulatory changes. For initiating the vision of the roadmap we recommend you to evaluate the trends on their unique potential. The major challenge is to extract those trends that provide a unique opportunity for the future of the business you are working[11].

> "Early identification and fast response to important trends and events which impact on the firm."
>
> IGOR ANSOFF[12]

Igor Ansoff, the eminent strategy theorist, proposed that strategic trend scanning should not only be based on extrapolations from statistical data, but also must rely on more immediate observations that provide initial and emerging indications of potential futures. He distinguished between strong signals and weak signals. Some trends come with strong signals and can be validated with data collection and statistical trend analysis. Such strong trends have long forecasting horizons of estimated impact, and therefore allow for building a knowledge base before executing an appropriate response. Other trends come with weak signals that appear more immediately in the present, as evolving signals[12]. According to Ansoff, weak signals are of equal importance in preparing a business for the impact of these trends.

Scanning weak signals involves heightened sensitivity and sustained reflection on the potential meaning of those signals. Intuitively sensing the importance of new observations requires professionals with an open-minded approach towards the environment. It is no wonder that designers and strategist adopted the STEP technique into their

STRATEGIC TREND SCANNING is the detection of change already underway in the environment of the organisation.

DESIGN ROADMAPPING

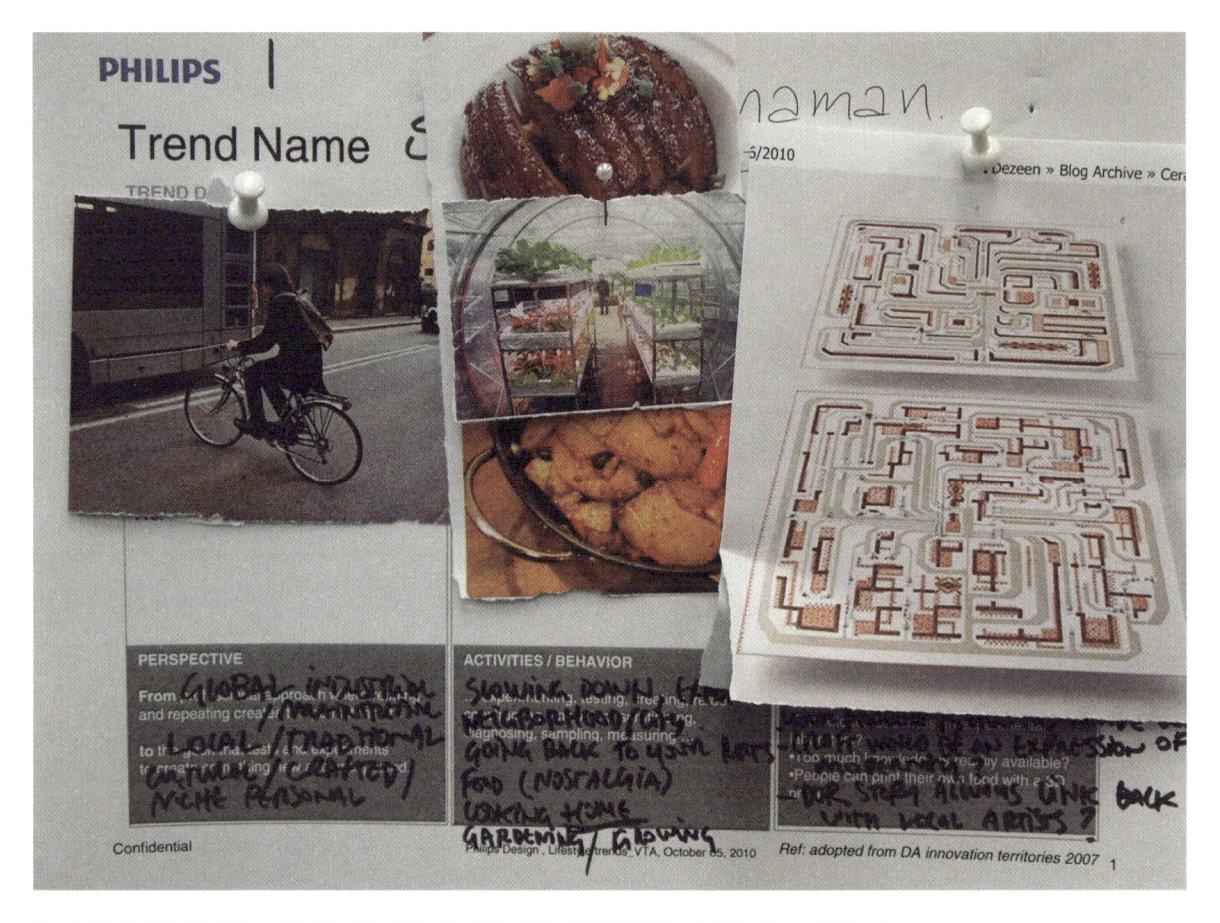

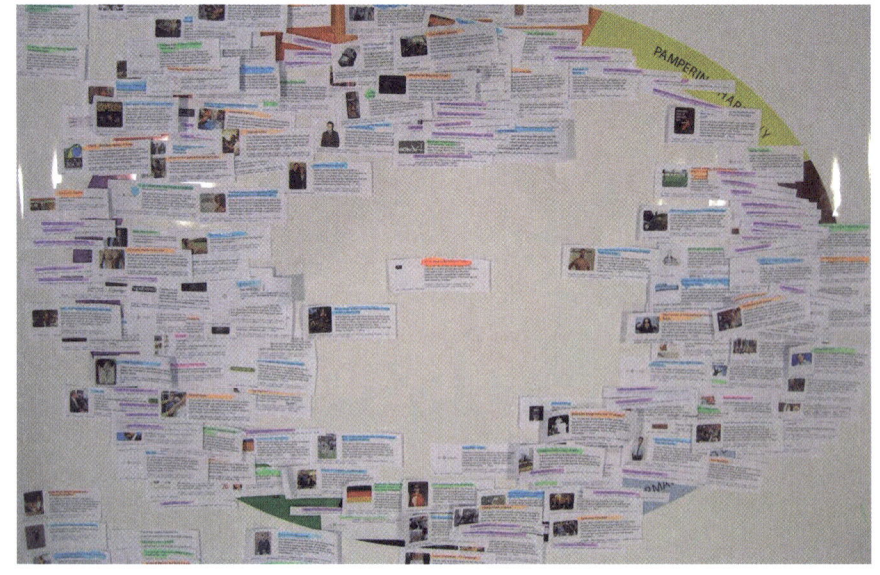

Creative Trend Research:
Trend patterns technique.

© Hanne Caspersen
photography,
PHILIPS DESIGN

Weak signal mapping in order to recognize patterns. To make relevant topic-specific trends we do this with weak signals derived from our network that has spotted these signals across different locations, based on our briefing - *Hanne Caspersen*.

trend research practice. In design practice we noted that designers also like to craft their own variants, such as the DESTEP technique, in which Demography has been differentiated from the Social domain and a new domain of Ecology has been added, reflecting the increased attention paid to sustainability.

One of the major challenges in trend scanning is to scan beyond the expected, and at the same time stay within the limits of understanding the complexity of the environment context. Researchers have found that the natural tendency of strategic managers is to scan narrowly, within the existing market. As a result, they fail to see competitive threats and innovation opportunities in the periphery of their current markets[13]. To overcome this pitfall you can help them to scan with an extended scope towards the edges. One place to look for breakthrough innovations is in fringe markets, such as for instance the snowboarding, microbreweries, and extreme sports once were. Nowadays these markets have become popular with wide audiences[13].

Another challenge is to scan trends related to people's behaviour, the so-called social-cultural trends. The behaviour of two groups of people is important. The first group comprises users and purchasers, and the second competitors and business partners. Scanners for trends within the first group often focus on user attitudes, activities, and interactions in a broader customer experience process. This scanning is aimed at uncovering patterns of user beliefs, wishes, and dilemmas. And in particular situations, it concentrates scanners' attention on trends in purchasing behaviour. Trend scanners of the second group of people in the industry and market environment predominantly addresses changes in the behaviour of competitors and business partners. This strategic scanning often makes use of self-established datasets, proprietary sources owned by market and sector research agencies, and large datasets established by a government-affiliated statistical institute (e.g., EUROSTAT, UNDP). In business, academically educated marketers and strategists prefer behaviour trends that can be validated by scientific methods. In strategic practice, advanced techniques of data mining expand the possibilities for this type of social-cultural trend forecasting.

Overall, we advise a combination of in-depth and statistical research in projecting rising or waning strategic trend patterns into the future. The DESTEP taxonomy technique is not only helpful in systematically checking the information gathered by desk research, but also enables brainstorming sessions on trends with designers, experts, and stakeholders. In such sessions, DESTEP is useful in making the (tacit) knowledge of the participating professionals explicit. By taking on a context view on the organisation, DESTEP stimulates the participants to grasp trends from six different angles. From each point of view, participants capture important signals of influence on the organisation's business.

Four Techniques for Creative Trend Research

LIANNE SIMONSE, NIYA STOIMENOVA & DIRK SNELDERS

We position creative trend research between the two extremes of visual trend spotting and strategic trend scanning and define creative trend research as: " the act of understanding in combining and unifying the isolated data of sensation into a recognizable whole of a trend". Designers can capture a combination of visual and textual signals by scanning the areas of interest, blending global and local trends, and blending visual and statistical trends. The result is an intuitive and creative type of trend research, which for the most part pertains to the behaviour of people doing their jobs and living their lives[14].

Figure 2.1 presents four techniques for creative trend research: trend scenarios, trend topics, trend views and trend patterns. All these creative techniques make use of trend synthesis. This is different from trend analysis; whereas an analysis breaks information into parts in order to identify causes and (trend) relations. A trend synthesis compiles information together in a creative way, by combining parts into new trend unities. The philosopher Kant has characterized synthesis as "the act of putting different representations together, and grasping what is manifold in them in one cognition - Kant[15]." The creative challenge in trend research is to unite in one sublime trend, the diverse mixed media forms of scanned bits and pieces.

We build the framework for creative trend research (figure 2.1) with at the heart two dimensions of (1) synthesis and (2) intuition that go hand in hand in carrying out creative trend research. Intuition allows the trend researcher, the person who will bring either sensed or memorized information together, to conceive a new trend. Intuition plays a vital role because a trend projects future value[16]. All these creative techniques are targeted towards identifying new opportunities for the creation of future value. Through intuition designers disclose the visible elements in connection to the sensed and memorized elements in trends. We introduce the four trend techniques below with examples from our readings in the literature, interviews, and work with students.

↘ Trend Scenarios

Generating trend scenarios has been described as an art, a craft, and a learning process, all implying the use of creativity[17,18]. Creating trend scenarios involves explorations that discover possible occurrences in the future [19]. "A trend scenario attempts to capture the richness and range of possibilities. At the same time, it organises those possibilities into narratives that are easier to grasp than great

Intuition and synthesis lie at the heart of CREATIVE TREND RESEARCH.

	VISUAL synthesis	MULTIMODAL synthesis
CREATIVE TREND RESEARCH		
SENSING intuition	Empathize emerging signals / Create Trend cards — **Trend Views**	Perceive frames of reference / Reperceive — **Trend Scenarios**
MEMORIZING intuition	Capture images / Visual clustering — **Trend Topics**	Immersion / Create trend patterns — **Trend Patterns**

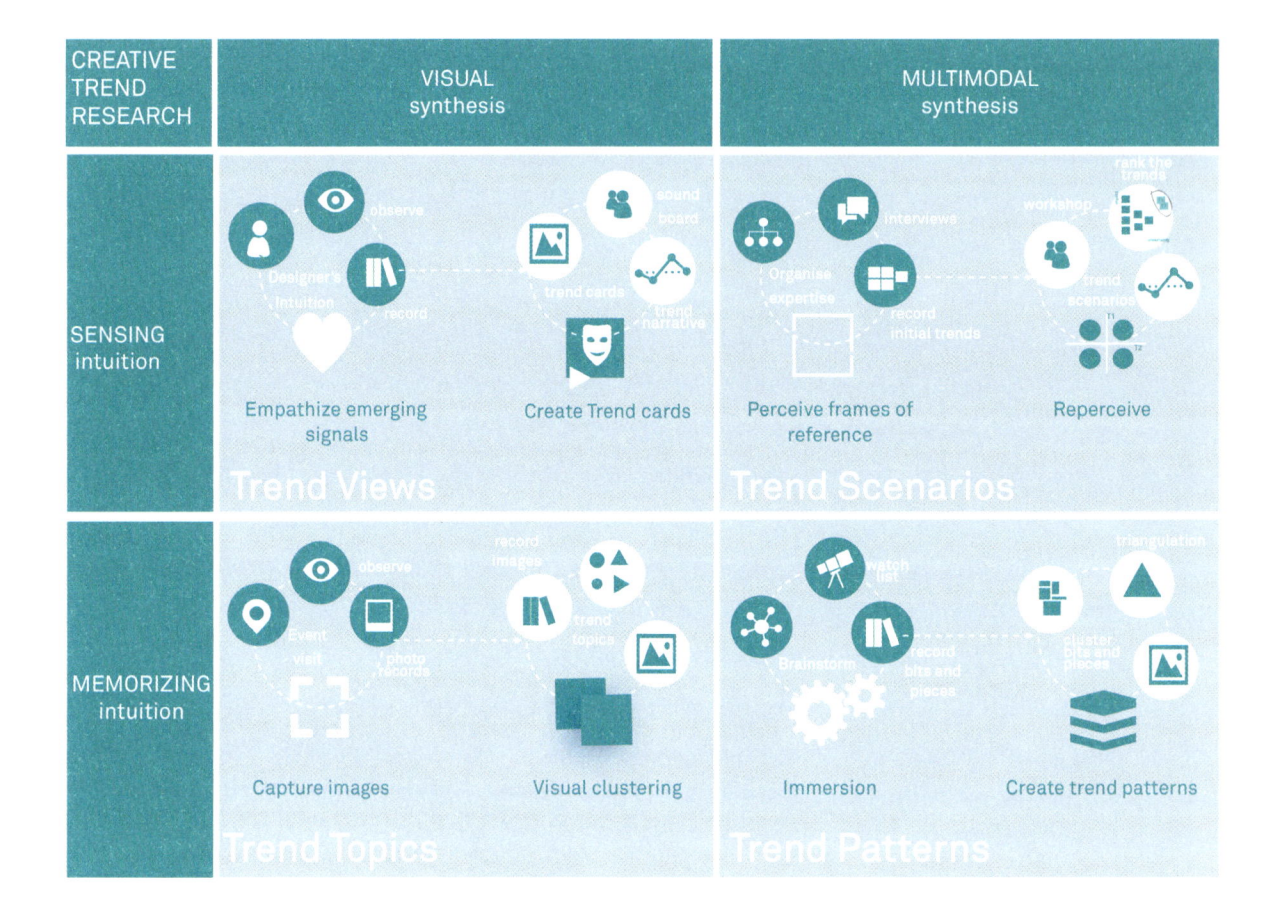

volumes of data[18]." In contemporary design practice, designers use visualisation tools to make trend scenarios come alive. They create storyboards and videos to tell a story about someone's future experience, making the imagined communicable. If done well, good stories become vivid in our minds, sometimes even more than our actual immediate reality.

According to Jan Nekkers, an highly experienced professional in trend scenarios, the important sources for scanning trends are knowledgeable people who perceive the future context from a personal view[20]. Before creating scenario storyboards the designer's job is to access the information that the involved persons sense and to collect multiple views on the future.

The two main activities of the trend scenarios technique are:

→ PERCEIVING: carry out several interviews with experts both from within and without the company. Important to ask is what people see as trends: "the trends people identify tell

←
Figure 2.1
Framework of creative trend techniques

cc Niya Stoimenova, 2016. Creative Trend Research: the role of methodology and intuition.
Paper for the master course Design Roadmapping. Faculty Industrial Design engineering, Delft University of Technology.

So, the trends are a mirror of their frames of reference. And maybe the trends people identify say more about themselves than about the reality -Jan Nekkers[20]." Besides uncovering and explicitly listing the most important trends for the future, a parallel intention is to discern the frames of reference of the involved decision-makers. In this way you can also uncover the dominant thinking inside an organisation. Scenarios can help people to come out of their frames of reference; the trend research also anticipates this by scanning trends that are opposed to, and break with, the dominant thinking patterns.

→ REPERCEIVING: the follow-up activity is to reperceive the environment in a workshop that brings the experts together. The workshop starts with a presentation of identified trends, followed by a brainstorm on additional trends. During the brainstorm, the DESTEP taxonomy is used as a checklist to draw attention to previously unconsidered domains. After this co-generation of trends, the workshop participants rank the trends on (user) impact and uncertainty, from high to low. The two trends with the most impact and uncertainty are selected as scenario drivers. By placing them as two opposing dimensions in a space, four quadrants arise out of which four scenario narratives are constructed.

The strongest benefit of having all the participants brainstorm and rank the trends together is that "they change the way they look to the business environment. They are reperceiving reality - Nekkers[20,17]." This collective and multimodal synthesis (framed in figure 2.1) also helps to develop a sense of ownership of the scenarios and hence increases further utilization of the trend results.

↘ Trend Topics

The technique of trend topics has been conceived at events such as the Première Vision in Paris, the leading trade fair in the fabric industry for apparel, which has a long history of influencing fashion. Today, trend topics have found their way into other industry events such as those for consumer electronics, aerospace, and interior design. 360-degree trend reports represent a notable example, comprising a visual selection of exhibition highlights from a specific industry, comparing emerging topics with previous years[21].

This trend technique relies heavily on visuals, mainly photographic material from the exhibition, clustered and thematised with the best representative images of each newly identified trend. We characterize the process of examining trend topics by the two activities of:

→
Urban Farming Trend

cc Trend view by blogger Amelys Erard, 2016.

→ CAPTURING: based on collecting visual information and taking photographs during the visits to influential events, such as trade fairs and design weeks. Suzanna Skalska, an experienced creative trend researcher, emphasizes that the activity of capturing is not a nine-to-five job: "it is 24/7 sensitivity to the world around[21]." It is also keeping up with several opinion-making magazines and TV programs, memorizing plenty of trend data regarding the industry under study.

→ CREATING TREND TOPICS: as the follow-up activity is based on visual clustering, of all gathered images from within and across events. The designer's intuition on the clustered visuals is crucial in creating trend topics. For example, Skalska visualised a trend topic called Trash Generation from a cluster of eleven images of emerging design work by young designers spotted at design weeks in different cities. She said, "For a few years now I have observed among a new generation of thinkers and designers a very big interest in trash. They all know that our daily rubbish will be a new source of material for future product development . . . this requires [a] new way of production as a challenge, but also as the only way to survive in the future. I call these new pioneers the Trash Generation - Skalska[21]." In synthesising such trend topics for an industry, designers often connect their findings both to local topics with short-term foresight, as well as to industry-transcending trend topics with long-term foresight.

Given that the designer has the creative lead to identify trends and create visual convincing topics, we framed the creative synthesis as a visual synthesis (see figure 2.1).

↘ Trend Views

Trend views are based on sensed information, but unlike trend scenarios, trend views are established by the single voice of an expert (see its framing in figure 2.1). Often this expert voice is that of a designer, who has sharply observed changes in the daily lives of people[22]. In trend views, the expert perceives signals directly, mostly as pictures, photographs and graphical images, which (s)he visually synthesises into mood boards, colour palettes, and other expressive visuals[23]. We distinguish two activities for the trend view technique:

→ SENSING AND RELATING TO THE SUBJECT: "the designer visits the places-where-it-happens to observe and become inspired, take pictures, make drawings, and discuss new developments with the people on location- Bol[22]." (S)he acquires additional trend knowledge through reading and design exercises such as making a mind map and a product lifecycle analysis. The designer interviews experts, for which (s)he might use pictures to trigger views and opinions about new trends. To check whether all areas are covered in this sensing exercise, (s)he would carry out a final DESTEP check. The designer then writes a short trend report, synthesising the most striking observations and insights, and her professional view on the most important upcoming trends. These trends are briefly summarized on trend cards - paper card, on which the trend is described in one or two sentences, an inspiring title, and a visual- [22].

→ CREATE TREND VIEWS: in this second activity, the designer maps the trend cards on a grid and decides on the two dimensions of the grid, clustering the trend cards into four scenarios. For each scenario, designers make a highly expressive visual or materialization (scale model, movie, etc.) to "feel connected to the subject as if one lives the scenario' and to delve into the details - Bol [22]." After designers develop the scenarios, a peer group discusses all the scenarios, sometimes with a client team, covering what is realistic and what one would want to avoid or stimulate. Next, the designer creates the trend views. (S)he formulates a design vision of the issues relevant to the future, including an agenda on what needs attention. Trend views present the possible, future experiences of possible, future users[22].

The discussion among designers in the second activity—create trend views—mainly seeks to strengthen the voice of the trend viewer. Other sources of creative authority (framed in figure 2.1) include formal design education from a respected school and a proven track record in research projects. In essence, trend views help designers and their clients empathize with people in a future situation.

↘ Trend Patterns

The creative challenge of the trend patterns technique is to spot robust substance from thin slices of emerging trends. According to the highly experienced trend researcher Hanne Caspersen, "trend research thrives on absorbed memories of prior experiences: through triggers of small pieces of signals, these memories can be retrieved and patterns recognized - Caspersen[24, 4]."

The trend patterns technique consists of two distinct activities: (A) intensive immersion into the context matter of the area of interest, and (B) a synthesis of trend patterns with authentication of different pieces of evidence.

→ IMMERSION, the first activity, is a deep dive into a specific area of interest to observe and absorb context. The creative lead starts out by mind-mapping the area of interest together with the stakeholders. As a result, (s)he makes a long list of their brain-dump to enable a better understanding of the subject matter and to apprehend what influences changes in relation to the topic. Next, the trend research team looks for clues and interesting bits of information collected through fieldwork, interviews, expert interviews, and "a lot of desk

→
Roadmapping
Trend patterns

© Bastian Schultes photography
PHILIPS DESIGN

Collaborative work-session where we are discussing the implications of the trends for future user experiences. From a series of workshops run simultaneously at different locations across the globe - *Hanne Caspersen.*

research, involving reading books and articles and searching on the Internet - Caspersen[24]." In-depth immersion requires the fieldwork of visiting people and having empathic conversations. Curiosity is critical in absorbing information and "you need to have a lot of respect for the subject matter - Caspersen[24]." Another part of immersion is a media 'watch list' compiled of Twitter accounts, webcasts, TV programs, events, experts, etc., which are all necessary as a foundation for further expanding the search and underpinning initial clues. This media investigation concentrates on observing and "absorbing the area of influence of opinion leaders and also alternative voices - Caspersen[24]." For memorizing, the results of all observations and mixed media bits are solidly documented for later use in evidencing trends.

→ CREATE PATTERNS : the second activity involves creating trend patterns in the absorbed information to uncover a new trend. When it becomes possible to connect and link several pieces into clusters and to discern the underlying meaning, a trend is conceived. In support of such a synthesis in trend patterns, (s)he uses mapping tools to uncover visual patterns in the mixed data. When creating visual patterns, you can use, for instance, a map layered with dimensions of people's motivations to ensure proper trend identification. Once an initial trend has been identified, it can be backed up with other slices of information gathered in the immersing step. The trends can also be linked to statistical data of social change to strengthen credibility

Through all of the confirmatory practices of the multimodal bits and pieces, the trend researcher triangulates the trend pattern in the second activity. After expansive research on bits of information that establishes a memory of the area of interest (as framed in figure 2.1).

Overall, we gained insights in four creative trend techniques. All these can be of important value to your efforts in formulating the user value drivers in design roadmapping. Future value drivers can be based on: Trend scenarios that include re-perceiving the organisation's mental frame of reference on value drivers (multimodal synthesis by the involved experts); Trend topics based on visual pattern creation of captured images; Trend views on foreseen trends (creatively sensed by the expert) or on Trend pattern creation from thin slices of documented memory (substantiated with multimodal evidence) that can provide a solid base for future value drivers.

PEERBY CASE

Shared Service Design
ANNA NOYONS

Inspired by the sharing economy trend, two entrepreneurs in Amsterdam started Peerby, a company that provides a peer-to-peer lending service. Via the online social media platform Peerby.com, users can borrow and share products with their neighbours. In this way, the utilization of products is extended. For example, an electric drill that is used by only one person is operated for 13 minutes on average during its lifespan, but can be put to more use when shared. More than 200,000 people have become members of this social community service and altogether they have registered over 1.5 million items that are available for sharing.

Besides saving money, time, and storage space, the users of the Peerby service also get to know their neighbours. Peerby's mission is realizing 'instant access to everything, everywhere and for everyone'. The entrepreneurs formulated their strategic identity by three pillars that they strongly believe in: social cohesion, sustainability, and social entrepreneurship. In addition, Anna Noyons, the chief product officer (CPO), commissioned an assignment to create a design roadmap that gains a longer-term perspective, supports the scaling up of innovation efforts and most importantly contributes to a positive member experience.

↘ ### Strategic trend scanning

A team of designers (all MSc candidates in strategic product design at Delft University of Technology) took up this challenge and created a roadmap for Peerby based on creative trend research. They explored that the origins of shared services, like Peerby's, thrive on experience value rather than ownership value. Building upon the popularity of sharing services such as Airbnb, and its frontrunner, couch surfing, various initiatives and new initiatives on product sharing pop up rapidly. The team of designers found twenty sharing services on the web in the Netherlands in 2013—and just one year later, in 2014, there were nearly two hundred. In other words, product sharing is booming, with successful businesses such as like Renttherunway (US) for sharing designer clothing and GetMyBoat for sharing a boat. The trend is almost becoming mainstream through this rapid increase of shared services business.

↘ ### Trend spotting on sharing services

For their trend spotting research the team gathered information from different sources, ranging from daily observations to global news magazines, from trend-watching blogs to a Forbes trend study and from

61   CREATIVE TREND RESEARCH

↑↗
Shared products service PEERBY members.

↑ PEERBY venture team.

© PEERBY photography.

collecting advertisements to a Harvard study on the subject of shared services. As guiding questions for their trends exploration they had formulated: "What triggers the interest in sharing?"
In a team meeting all trend data was brought to the table and each element was discussed in terms of user impact and their relevance for the Peerby service. On the wall, they clustered the relevant findings (dropping the irrelevant and those with almost no user impact) and created patterns of substantial visual and textual data. For each pattern, they formulated a trend theme.

Trend Roadmap

↙ Figure 2.2
Trend roadmap for PEERBY

cc Tess Poot, Tonino Gatti, Lars Scholings, Eva van Genuchten & Pepijn van der Zanden. PEERBY Project Report Design Roadmapping Master Course, Faculty Industrial Design Engineering,
Delft University of Technology.

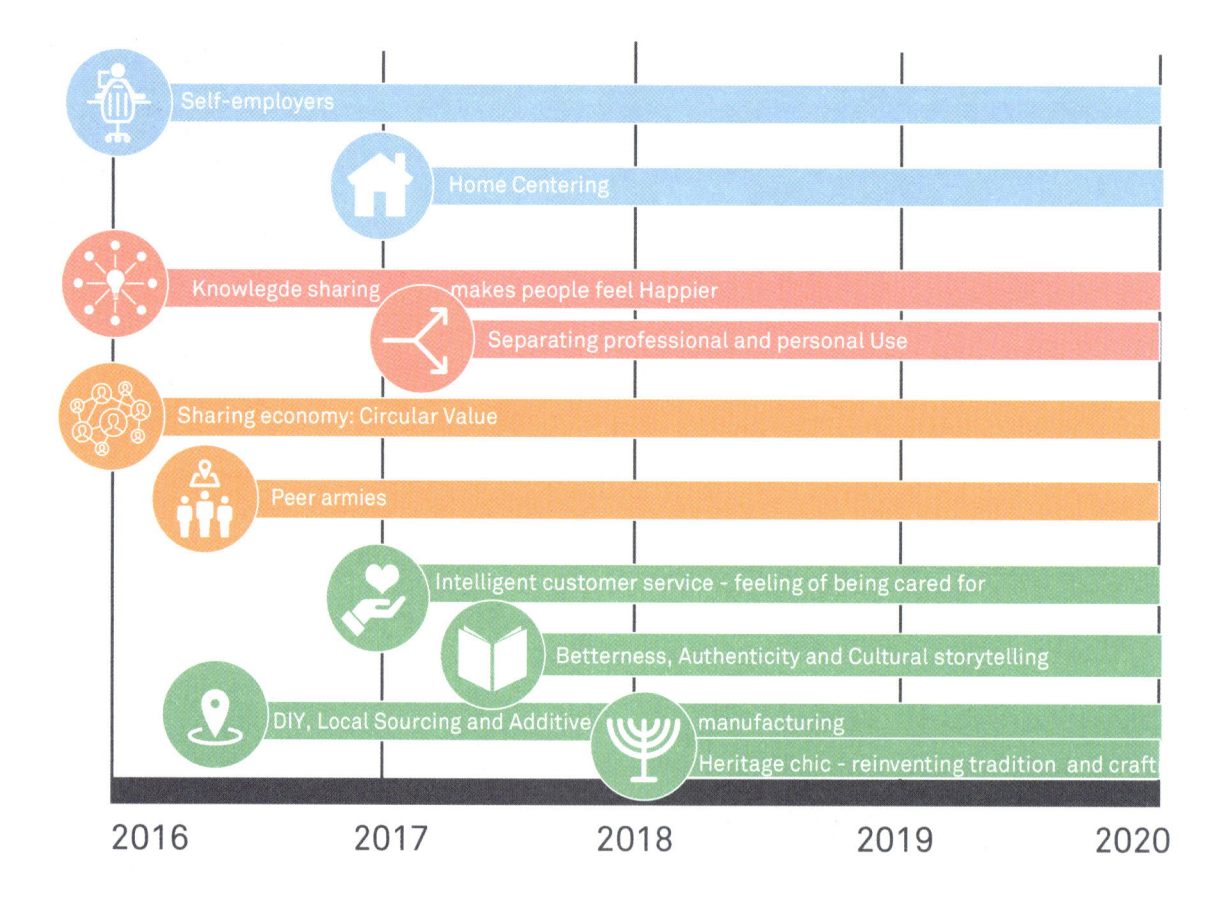

↘ Top 10 trends listing

- Self-employers
- Home Centering
- Knowlegde sharing makes people feel Happier
- Separating professional and personal Use
- Sharing economy: Circular Value
- Peer armies
- Intelligent customer service - feeling of being cared for
- Betterness, Authenticity and Cultural storytelling
- DIY, Local Sourcing and Additive manufacturing
- Heritage chic - reinventing tradition and craft

2016   2017   2018   2019   2020

→ SELF-EMPLOYERS
The boundaries between work and private life will continue to blur; social status will replace work status, and more people will become self-employed. This is unleashing creativity. More than a million "makers" are currently selling their own designed products via online marketplaces like Etsy, and as many people have become "hosts" on short-term accommodation platforms like Airbnb. The home is becoming the center for all kinds of different activities. Working and living will take place in the same spot. The number of self-employed people in the US grew by one million. Work and life will become further intertwined.

→ **HOME CENTERING: HYPER-EFFICIENCY OF INSTANT ACCESS**
Consumers are seeking faster and more efficient ways to overcome issues of lack of time, space and resources. They want instant access to products, even faster than going to the shop or hiring it. Additional match and bring services are driving service providers to the limits.

→ **KNOWLEDGE SHARING MAKES PEOPLE FEEL HAPPIER**
Harvard University has studied the experience of sharing and giving. The researchers found that sharing activates those parts of the brain that relate to feelings of happiness. This indicates that sharing makes people feel happier.

→ **SEPARATING PROFESSIONAL AND PERSONAL USE**
User-generated boosting by uploading photos and video, writing blogs and posting comments, users are actively generating, updating and creating content as well. One retailer claims that online shoppers who come from Pinterest are three times more likely to make a purchase than the average shopper on their site. At best this boosts the online reinforcement of "Keeping Up with the Joneses."

→ **PEER ARMIES**
After "I", the era of "we" is emerging. After that advertisements focused dominantly on the use of 'i' to denote individuality (iphone, I amsterdam, etc.), highlighting in particular the product properties that enhance personal identity, a countertrend has been observed. Under the influence of social media, a growing number of advertisements show groups of friends announcing the era of "we": an era in which personal development and reputation depend largely on communities of friends.

→ **SHARING ECONOMY: CIRCULAR VALUE**
The circular economy aims to generate more value and economic opportunity with less material and energy consumption. Through the power of circling, materials are kept longer in use by multiple cycling or by lengthening cycling duration. Creating value exchanges instead of transactions also includes using "alternative currencies" such as time, reputation and craftsmanship products. This business model innovation can provide new growth opportunities for material, component and product reuse.

→ **INTELLIGENT CUSTOMER SERVICE – FEELING OF BEING CARED FOR**
Open access services are further challenged by artificial intelligence technology that will make it possible to "read" consumers and give them what they want, even before they ask for it.

→ **BETTERNESS, AUTHENTICITY AND CULTURAL STORYTELLING**
In a world full of buzz and superficial interactions, people are seeking more authenticity, depth and meaning. They choose to be offline and use their leisure time for cultural experiences, creative expressions and storytelling. These activities are all about mutually beneficial relationships. They are intended not only to make those who engage in them better people, but in a way that makes these people part of a larger collective – social actualisation.

→ **DIY, LOCAL SOURCING AND ADDITIVE MANUFACTURING**
More local and do-it-yourself production will be made possible by additive manufacturing facilities such as 3D scanning and 3D printing. Co-creation and partly personalized products become more feasible and viable to ensure perfect fit.

→ **HERITAGE CHIC – REINVENTING TRADITION AND CRAFTSMANSHIP**
In both Europe and Asia, cultural traditions such as beer brewing or scarf knitting are being preserved, but also reinvented – and often by the people you'd least expect: serve, facilitate and connect with them!

↘ ## Design Roadmap for PEERBY

The design team shortlisted 10 trends (see figure 2.2) and ranked them on long term user impact and how Peerby could respond with innovations (Innovation strategy fit). This provided inspiration and direction for formulating the roadmap vision: 'Circular community value' based on the highest ranked trend: "Through the power of circling the products in a neighbour community, they are kept longer in use by multiple cycling and by lengthening cycling duration". Then they decided on which trends to map on the market roadmap with a view to providing the most unique and promising values for Peerby:

→ Betterness, authenticity, and cultural storytelling
→ DIY, local sourcing and additive manufacturing
→ Intelligent customer service – feeling of being cared for
→ Heritage chic—reinventing tradition and craftsmanship

In creating the roadmap, the team mapped these trends to the timeline. Figure 2.3 presents the roadmap in connection to the trends. They shaped the future direction, building upon the shared neighbourhood service of lending and borrowing stuff, they could also create things together. For instance, thanks to help from a neighbour, one participant was able to build a table. They propose that Peerby should grow and become a truly established community platform by emphasizing more the interactions between neighbours and building their services around environmental awareness in combination with the momentum of the sharing-economy. To communicate this, the design team made a roadmap that contains four horizons over a period of ten years.

CREATIVE TREND RESEARCH

↘
Top trends combination: Repair Café event.

An event where volunteers with repair skills invite members of the community to bring their broken things in to be fixed. They work together to understand the problem and find a solution. The fixers share their knowledge and tools to show the owners of the items how to do the repair.

cc Repair Café Toronto, photography.

ANNA NOYONS is a strategic product- and service designer. After working as a design consultant for several years for big and small companies, she joined the PEERBY team in a very early stage as the Chief Product Officer and helps to create the companies vision and strategy and translate that into the product experience.

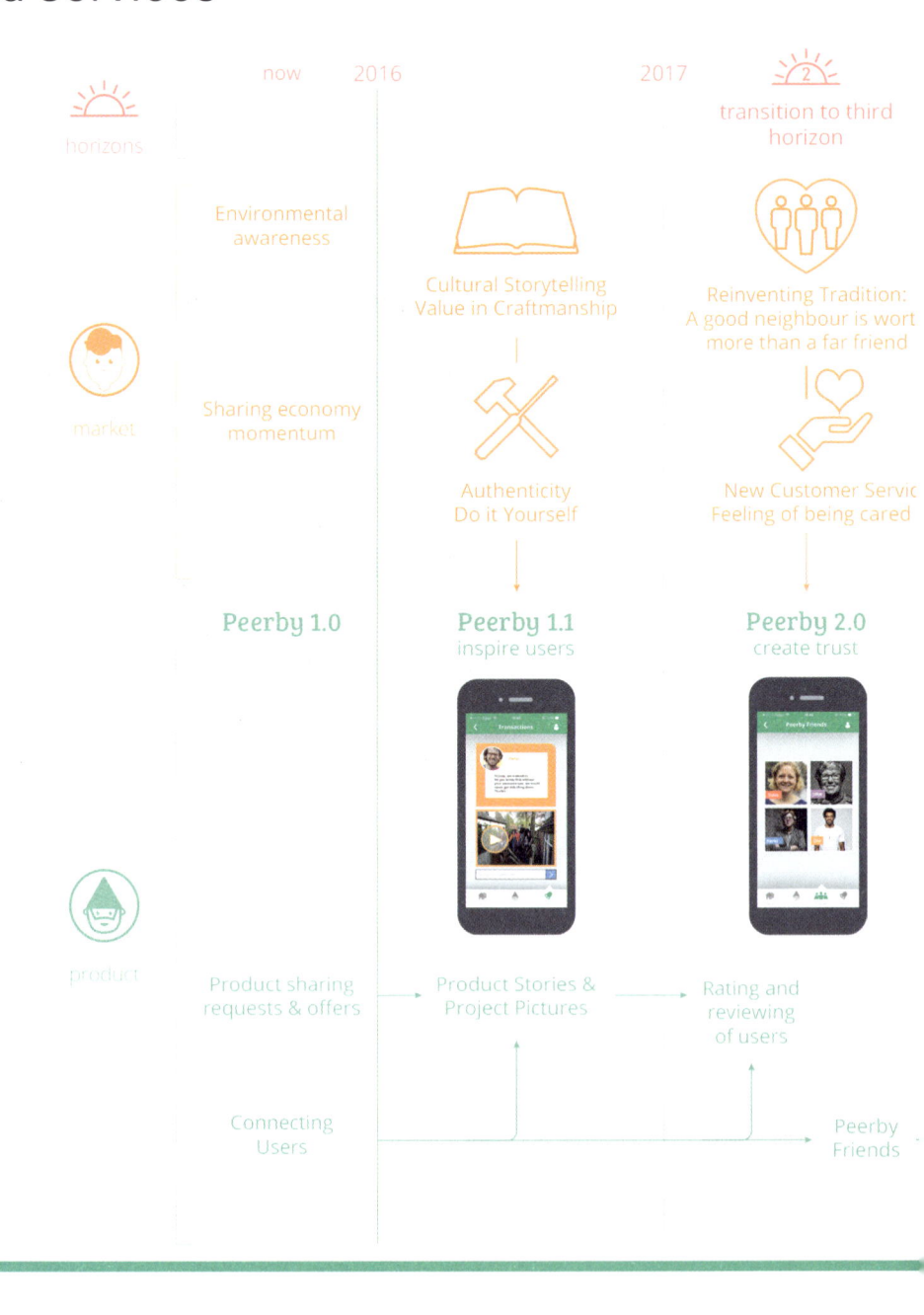

DESIGN ROADMAP
PEERBY Shared services

horizons
- now | 2016 | 2017 | transition to third horizon

market
- Environmental awareness
- Cultural Storytelling / Value in Craftmanship
- Reinventing Tradition: A good neighbour is worth more than a far friend
- Sharing economy momentum
- Authenticity / Do it Yourself
- New Customer Service / Feeling of being cared

product
- Peerby 1.0
- Peerby 1.1 — inspire users
- Peerby 2.0 — create trust

- Product sharing requests & offers
- Product Stories & Project Pictures
- Rating and reviewing of users
- Connecting Users
- Peerby Friends

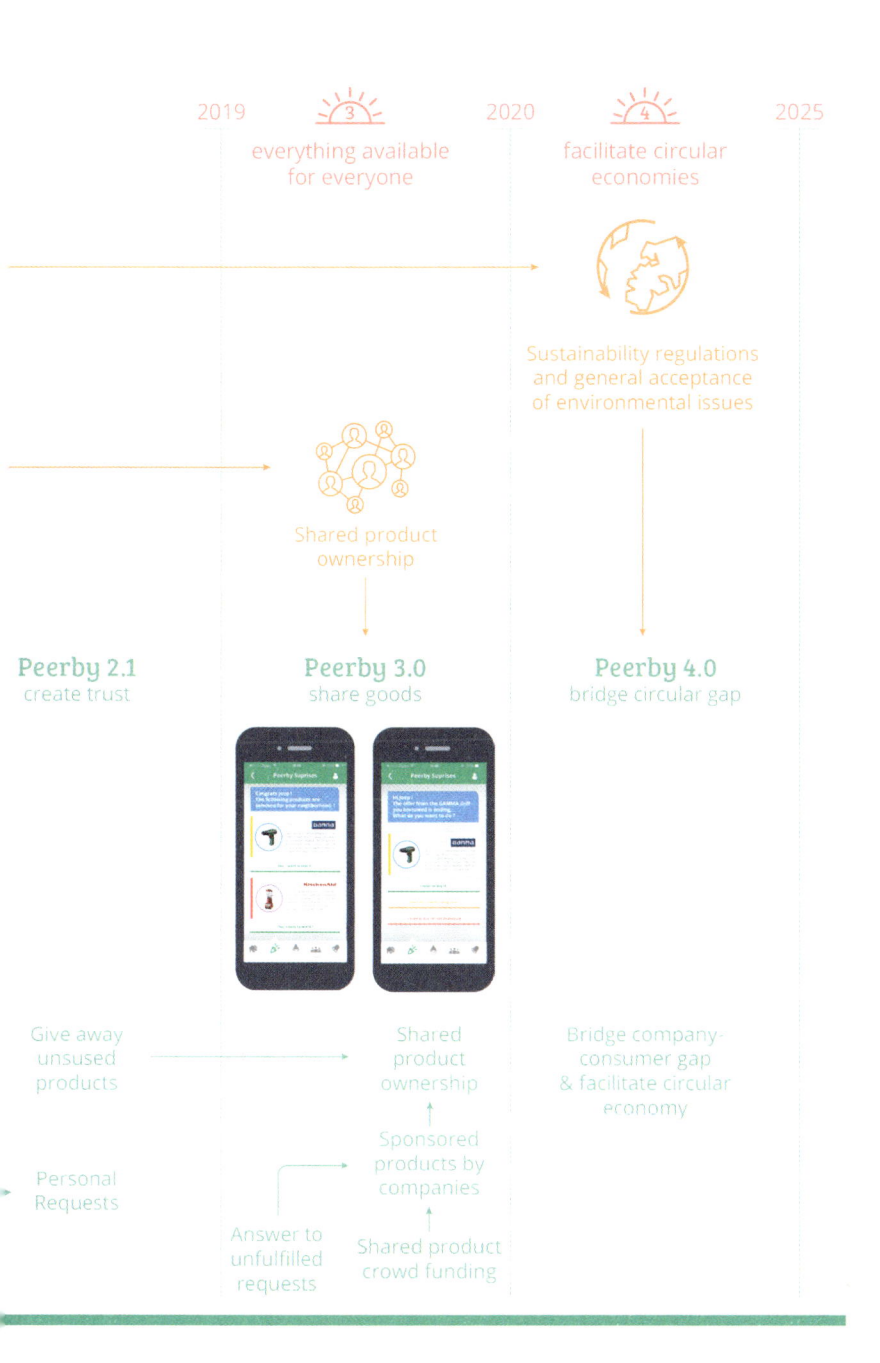

Figure 2.3
Design roadmap for PEERBY

cc Tess Poot, Tonino Gatti, Lars Scholings, Eva van Genuchten & Pepijn van der Zanden. PEERBY Project Report Design Roadmapping Master Course, Faculty Industrial Design Engineering,
Delft University of Technology.

Please note that the design roadmap is created for PEERBY by Strategic Product Design Master students, and therefore do not reflect Peerby's actual strategy.

LAB ↗

Try out the Trend Topics technique

MATERIALS NEEDED:
→ lifestyle magazine
→ access to the Internet to search the web for images

1 Take a lifestyle magazine or decide to spend two hours on the Internet and use a search engine with an image view (for instance Google Image).

2 Formulate an area of interest, a 'radar' to start your trend research, for instance, a lifestyle activity (e.g., dining, sporting, etc.), an industry (cars, drones, retail), or a product/market/technology combination (coffee machines, 3D printing, etc.).

3 Capture images that are new and innovative and in some way express a promise for the future. Go for quantity and collect about fifty images.

4 Cluster the images that have similar elements and label these elements. Identify trends with a second round of clustering the image clusters. Create larger clusters that connect or unify the image clusters to each other and propose a term for the trend cluster.

5 Arrange the trend clusters on the decision grid of user impact and innovation fit (see figure 6.5). Rank the trends from high to low impact and high to low innovation urgency.

6 Choose the meaningful trends with significant to high impact on user values for your top 10 listing of trend topics.

7 For each trend topic, create an inspirational title, a few sentences of explanation, and one characterising image.

This kind of try out you could easily do in one sitting, usually in less than four hours if you're taking this simple, hands-on trend technique. It's a perfect opportunity to try out your creative trend research skills.

Prof. dr. SUSAN REID on the subject of Market Visioning and Creative Trend Research

LS In design roadmapping, formulating a market vision is an essential element. Your scientific research on market visioning[11,25] shows that for successful development of radically new, high-tech products, a market vision is a prerequisite. How do the firms you researched start market visioning?

SR Well, according to my research, market visioning competence and market vision each have separate and cumulative impacts on early performance of firms involved with radically new high-tech products. The inherent risks and rewards associated with these types of high-stake ventures require that firms create long-term visions to guide their efforts. A clear and compelling vision about the product-market opportunities associated with radically-new ventures can help firms to achieve superior competitive advantages.

LS Have you come across the use of (creative) trend research used in formulating a market vision?

SR Firms can come to a market vision from many different paths, including those routes to creating a market vision, which is largely top-down versus those which are more bottom-up. It would be more likely the case that if a firm is 'cognizant' of formulating a market vision and actively striving to do so that it would be a case of top-down formulation. In such cases, then, trend research might be used to create the market vision.

LS What do you consider as the most important elements in creating a market vision?

SR The numbers would say that the strongest, most important component of market visioning competence (in terms of statistical impact) was idea-driving, championing behaviour on the individual side, and having a proactive market orientation on the organisational side.

LS In your research, you showed that the impact of market visioning on innovation performance is positive, in particular on the key aspects of the early performance, early success with customers, and ability to attract capital for the business development plans behind the radical innovations. What are your thoughts about the impact of creative trend research?

SR I tested the model as a whole and found that market learning tools, which would encompass techniques like trend research, is a statistically significant component of the market visioning competence, which has an impact directly on ability to attract capital and indirectly on early success with customers. Importantly, I used four items to measure market learning tools and the item which had the highest standardized loading and reliability was measured using the item: "we tried to develop several potential technological scenarios before choosing market(s) to pursue"[26]. This leads me to believe that in fields governed by technological development, it's important for the researchers to focus on technology trends in addition to the market trends in their field, and to do so early on, in the front end of innovation.

LS What are your recommendations for design roadmappers on creating a market vision?

SR I think it is important to try and understand the context under which a vision is being 'born', both in terms of initiation with a low or high level of market contex, and in terms of whether the focus is from an individual or organisational perspective[27]. The easiest to map will be in the case of an innovation being initiated from an organisation and having high extend market context. To this end, design roadmappers might start to work on contextualizing a given market vision form.

SUSAN REID is a professor at Bishop's University, teaching and researching in areas where marketing, innovation and entrepreneurship intersect: market vision, networks, radical innovation, brand and innovation management. She has published in refereed journals including: Journal of Product Innovation Management, Technological Forecasting and Social Change, R&D Management, World Development, Business History and International Journal of Technology Marketing. Her academic background combines with over 25 years of consulting, business and board experience for the aviation, biopharmaceutical, nanotechnology and consumer goods sectors (including for the ice cider and spirits business, Domaine Pinnacle, co-founded with her husband Charles Crawford in 2000 and sold in 2016). Her mission is to help others negotiate the path between their passion and successful product/market vision.

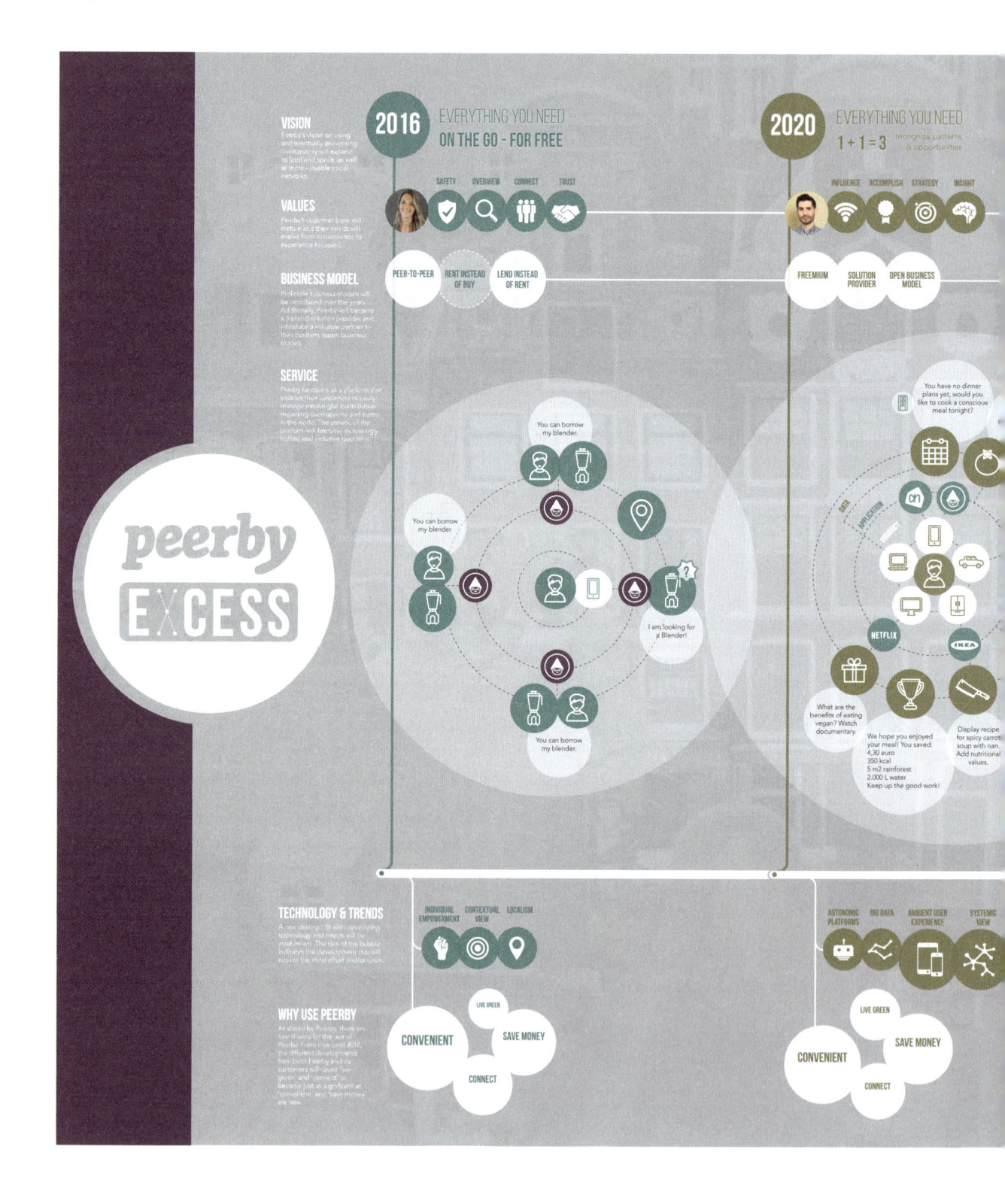

DESIGN ROADMAPPING

CREATIVE TREND RESEARCH

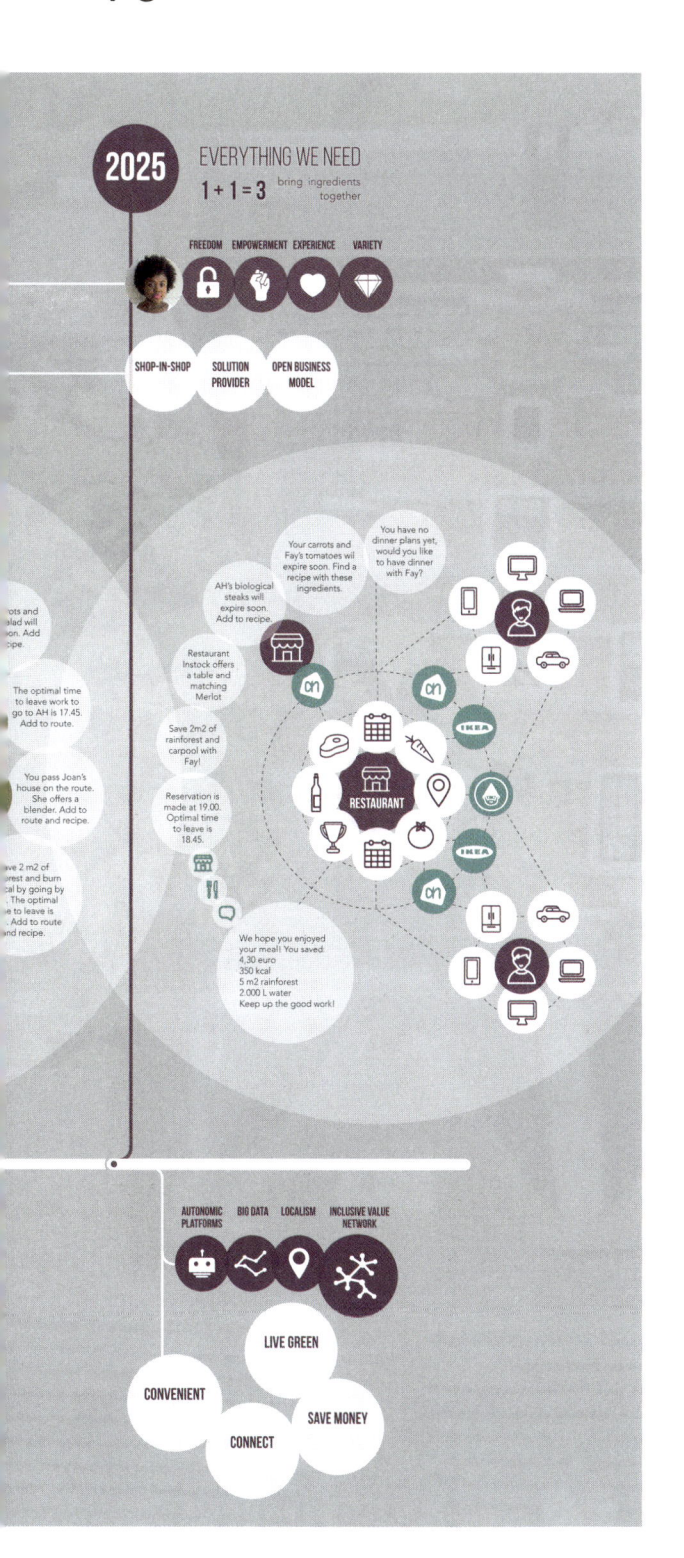

Design roadmap for PEERBY, Jan Buijs award, 2016.

cc Esmee Mankers, Mark Kwanten, Nienke Nijholt & Ben Hup.
PEERBY Project Report Design Roadmapping Master Course, Faculty Industrial Design Engineering, Delft University of Technology.

Please note that the design roadmap is created for PEERBY by Strategic Product Design Master students, and therefore do not reflect Peerby's actual strategy.

↑
Shared Products
services
PEERBY members

© PEERBY photography.

IN SUM

In this chapter, we positioned creative trend research between the two extremes of visual trend spotting and strategic trend scanning. We gained the insight that creative synthesis rooted in human intuition is the key characteristic of creative trend research. Creative synthesis unites bits and pieces of mixed media into a trend creation that imagines future value.

Creative synthesis can be established by:
- → Designer's view on foreseen trends (creative authority of the expert).
- → Visual pattern creation of captured images (peer-to-peer authority).
- → Re-perceiving the organisation's mental frame of reference (collective authentication of trends by the involved experts).
- → Pattern creation from thin slices of documented memory (substantiated with multiple evidence authentication).

The synthesis is driven by creative authority or authentication of multiple sources.

With a framework four creative techniques are distinguished: trend views, trend scenarios, trend topics, and trend patterns on the dimension of intuition and creative synthesis. The Peerby case demonstrated the importance of trend research for the roadmap. The lab provides a do it yourself experience on the art of trend research.

---

1. Evans, M. (2011). Empathizing with the Future: Creating next-next generation products and services. The Design Journal, 14(2), 231-251.
2. Oxford English Dictionary Online, accessed June 2016.
3. Reynolds, W. H. (1968). Cars and clothing: understanding fashion trends. Journal of Marketing, 32(3), 44-49.
4. Gladwell, M. (2007). Blink: The power of thinking without thinking. UK: Little, Brown and Company, Hachette Book Group.
5. Philips Design Stories: People centric research: Putting people at the heart of innovation. https://www.90yearsofdesign.philips.com/article/87, accessed August 2016.
6. Gaimster, J. (2012). The changing landscape of fashion forecasting. International Journal of Fashion Design, Technology and Education, 5(3), 169-178.
7. Goldsmith, R. E., & Clark, R. A. (2008). An analysis of factors affecting fashion opinion leadership and fashion opinion seeking. Journal of Fashion Marketing and Management: An International Journal, 12(3), 308-322.
8. Rickman, A.T. & Cosenza, R.M. (2007). The changing digital dynamics of multichannel marketing: The feasibility of the weblog: text mining approach for fast fashion trending. Journal of Fashion Marketing and Management: An International Journal, 11(4), 604-621.
9. Boyd, T. J., Okleshen Peters, C. & Tolson, H. (2007). An exploratory investigation of the virtual community MySpace.com: What are consumers saying about fashion? Journal of Fashion Marketing and Management: An International Journal, 11(4), 587-603.
10. Aguilar, F.J. (1967). Scanning the business environment. New York, NY: Macmillan Co.
11. Reid, S. E. & Brentani, U. de. (2012). Market Vision and the front end of NPD for radical innovation: The impact of moderating effects. Journal of Product Innovation Management, 29(S1): 124–139.
12. Ansoff, H.I. (1980). Strategic issue management. Strategic management journal, 1(2), 131-148.
13. Day, G.S. & Schoemaker, P.J.H. (2004). Driving through the fog: Managing at the edge. Long Range Planning, 37(2), 127–142.
14. Tovey, M. (1997). Styling and design: Intuition and analysis in industrial design. Design Studies, 18(1), 5-31.
15. Kant, E. (2001). Critique de la raison pure (1781). PUF, 63-64.
16. Badke-Schaub, P. & Eris, O. (2014). A theoretical approach to intuition in design: Does design methodology need to account for unconscious processes? In: An Anthology of Theories and Models of Design (pp. 353-370). London: Springer.
17. Wack, P. (1984). Scenarios: The gentle art of re-perceiving: A thing or two learned while developing planning scenarios for Royal Dutch-Shell. Cambridge: Harvard Business School, division of Research.
18. Van der Heijden, K., Bradfield, R., Burt, G. Cains, G. & Wright, G. (2002). The Sixt sense: Accelerating organisational learning with scenarios. Chicester: Wiley.
19. Nekkers, J. (2016). Developing Scenarios. In: Foresight in Organisations. Methods and Tools (pp. 11-40). London: Routledge.
20. Nekkers, J. (2016) on Creative Trend Research: the role of methodology and intuition. Transcripts by Stoimenova, N. (2016).
21. Skalska, Z. (2016). Internet blog http://www.360inspiration.nl/inspiration/trash-generation/ accessed August 2016.
22. Bol, S. (2016) on Creative Trend Research: the role of methodology and intuition. Transcripts by Stoimenova, N. (2016).
23. Bol, S. (2016). Creativity in foresight: Seven exercises. In: Foresight in Organisations. Methods and Tools (pp. 190-200). London: Routledge.
24. Caspersen, H. (2016). on Creative Trend Research: the role of methodology and intuition. Transcripts by Stoimenova, N. (2016).
25. Reid, S. E. & Brentani, U. de. (2010). Market Vision and Market Visioning Competence: Impact on Early Performance for Radically New, High-Tech Products. Journal of Product Innovation Management, 27(4): 500–518.
26. Reid, S.E., Roberts, D. & Moore, K. (2014). Technology Vision for Radical Innovation and its Impact on Early Success. Journal of Product Innovation Management 32(4): 593–609.
27. Reid. S. (2015). Vision and Radical Innovation: A Typology. In: Adoption of Innovation. Switzerland: Springer International.

FUTURE VISIONING

FUTURE VISIONING

HOW TO DESIGN A FUTURE VISION

The leading question for designers to start-up a roadmapping process is: what is our future vision, because that vision establishes the destination for the roadmap. Before designing anything, it is important to discern its properties. Therefore, in this chapter we first take a deep-dive into the properties and definition of a future vision. - what it is and what it is not. After that, we elaborate on special kinds of future visions — ones that lead to inspirational artifacts like concept cars and concept kitchens. These are in particular designed for the future exploration of and communication about disruptive value innovations. In roadmapping, we consider such vision concept as the ultimate future vision – Then, we discuss the role of designers in taking the 'creative lead' in formulating the future vision for the organisation's innovation strategy. On top of that we present a special case story of future visioning in design practice. Flavio Manzoni, Senior Vice President of Design at Ferrari shares his notions on taking inspiration from the future rather than building from the constraints and legacy of the past.

Future Vision

↘ Creatively expressing a desired future

On a design roadmap, the future vision points to the destination. As an expression of a desired future[1,2,3,4], the vision provides a strategic reference point — a focused direction that leads to stronger motivation[5]. Visions imagine experiences of future innovations.

A vision's creative expression may be plain or elaborate, down-to-earth and practical or dreamily utopian, and can come in the form of graphic, video, artifact or written narratives[6]. Creative expression makes the vision more explicit; in order to inform actions in concordance with it until the vision is either achieved or replaced. Unlike a goal, a vision aims to establish a tension between "what is" and "what could be" [7], so as to provide direction for the innovations on the roadmap that lead to it.

↘ Imagining a desired state of the future: capture value wishes

A future vision imagines possible future experiences concretely. It listens to people's wishes for future innovations and articulates them explicitly[8]. A future vision holds imaginative, and dreams about the future. It can only take root through true leaps of inspiration, which are sometimes based on observed trends and identified opportunities, and sometimes on personal inspiration or intuition. As Carl Jung, once said, "Your vision will become clear only when you can look into your own heart . . .who looks outside, dreams; who looks inside, awakes[4]." Visioning is often seen as the realm of the artist, the poet, the tinkerer, the futurist and the designer[4].

Capturing wishes about the future as people would like it to be is the art of visioning. People's passions, desires and aspirations can be

FUTURE VISIONING

A FUTURE VISION is an expression of a desired future.

framed as future visions. For instance, a film fan who's "wish is to watch a movie that feels like being there" inspired the vision driving a multimedia venture roadmap: make immersive movie experiences happen. The value propositions for such experiences do not exist yet. The vision captures a value wish[8]. Value wishes express a desired end-state in which a novel value fulfils an unmet need or resolves a present dilemma or feeling of frustration experienced by a user target group. Besides wishes for new products, people's value wishes might be to alleviate the effort that goes into annoying or "dirty" jobs, or perhaps reduce risk and feel safer[8].

↘ The strategic reference point for actionable innovations

Visions also contain a specific call to action. "Dreamers dream about things being different. Visionaries envision themselves making the difference[9]." The vision not only imagines what needs to be different in the future – it expresses specific, achievable dimensions to making the difference. As Alan Turing said, "We can only see a short distance ahead, but we can see plenty there that needs to be done[10]." Many roadmappers consider it imperative to develop a vision that includes a specific desire for action as the vision channels the energy of the different innovation professionals into useful actions and activities of research, design and innovation[11,8].

Visioning entails taking a leap toward the future[12]. To get there, some innovation professionals may want to take high risks, while others might prefer a more risk-averse direction. In the theory on creating a strategic reference point, scholars have framed that these two types of risk behaviors meet somewhere in the middle. Perceived gains and losses from both perspectives at the end of the risk spectrum that are voiced, discussed and reviewed in strategic dialogues often lead to finding an optimal vision that lies in between – the so-called 'strategic point of reference'[5]. Therefore it is important to organise a team of innovation professionals to create a vision together. Through the sharing of personal views on the future and evaluating the imagined opportunities from multiple perspectives, the future will no longer stay in the domain of individual knowledge – it becomes one of collective action[8].

Overall, in roadmapping, the more clear the strategic reference point of the vision is, the more easily everyone can see how to build a path towards it.

↘ Four distinguished properties

Beyond bringing the imagination and realisation contributions of the innovation professionals together into one strategic point of reference – the design of the vision expression provides the creative challenge. According to research, strong visions have four distinguished properties of clarity, value drivers, artifact and magnetism.

- → CLARITY in the vision expression enables immediate understanding of what it would be like to experience the future innovation in the explicitly expressed desired end state[13,16,25].
- → VALUE drivers capture the key compelling benefits of value wishes: wherein the specific value fulfils an unmet need or solves a dilemma of a user target group in the future[8,14].
- → ARTIFACT, materialise the imagined value wishes with images in 2D or 3D- dimensions[15].
- → MAGNETISM involves the desirability and attractiveness of the vision – 'the thing' the vision creators are truly passionate about in such a way that it potentially energizes others to direct their actions towards it[16].

A future vision is not a design vision

Design, as a discipline, has always been closely linked to the exploration of the future[17]. Depending on the design task, to varying extents designers act as "futurists or futurologists, in the field"[6]. In certain product design approaches, such as 'Vision in Product', designers are encouraged to formulate a personal vision on the direction of the solution design[18]. This design vision is however, mostly formulated by the designer in an independent way, grounded in his or her design authority and craftsmanship in creating solutions.

In contrast, a roadmap vision is in principle a shared vision, co-created by a number of innovators including marketers, engineers and users involved[11,8]. Moreover, a roadmap's vision is created from the future intent[8]. This vision explores the future, whereas a design vision is created after design research on certain problems in the present[15].

A future vision is not a corporate vision

Because a roadmap's vision is focused on future innovations, its target and scope differ from the mission and vision statements found in many corporate strategies. These organisation visions have a larger, companywide scope, covering the raison d'être, overall positioning and goal setting of the company[19]. Corporate visions – examples are readily available on the Internet - see for instance the one of Apple– include strategic statements related to the strengths of the company, and its corporate values and beliefs[20].

A roadmap's vision is different due to its particular focus on innovation and future value experiences[12]. That being said, corporate visions and the future visions of roadmaps can be related – both top down and bottom up, a future vision can be embedded in the corporate vision.

Lift, 2000

© Fiona Tan.
screenprint, 108 x 64 cm
Courtesy the artist and Frith Street Gallery, London.

In the morning of 14 January, 2000, Dutch artist Fiona Tan gathered fifty large red helium-filled balloons, affixed their strings to a harness and was briefly lifted into the chilly winter air above Amsterdam's Sarphati park. She drew the attention of passers-by walking their dogs and their children. Tan: "This was a dream I have had for a long time, ever since childhood, in fact."

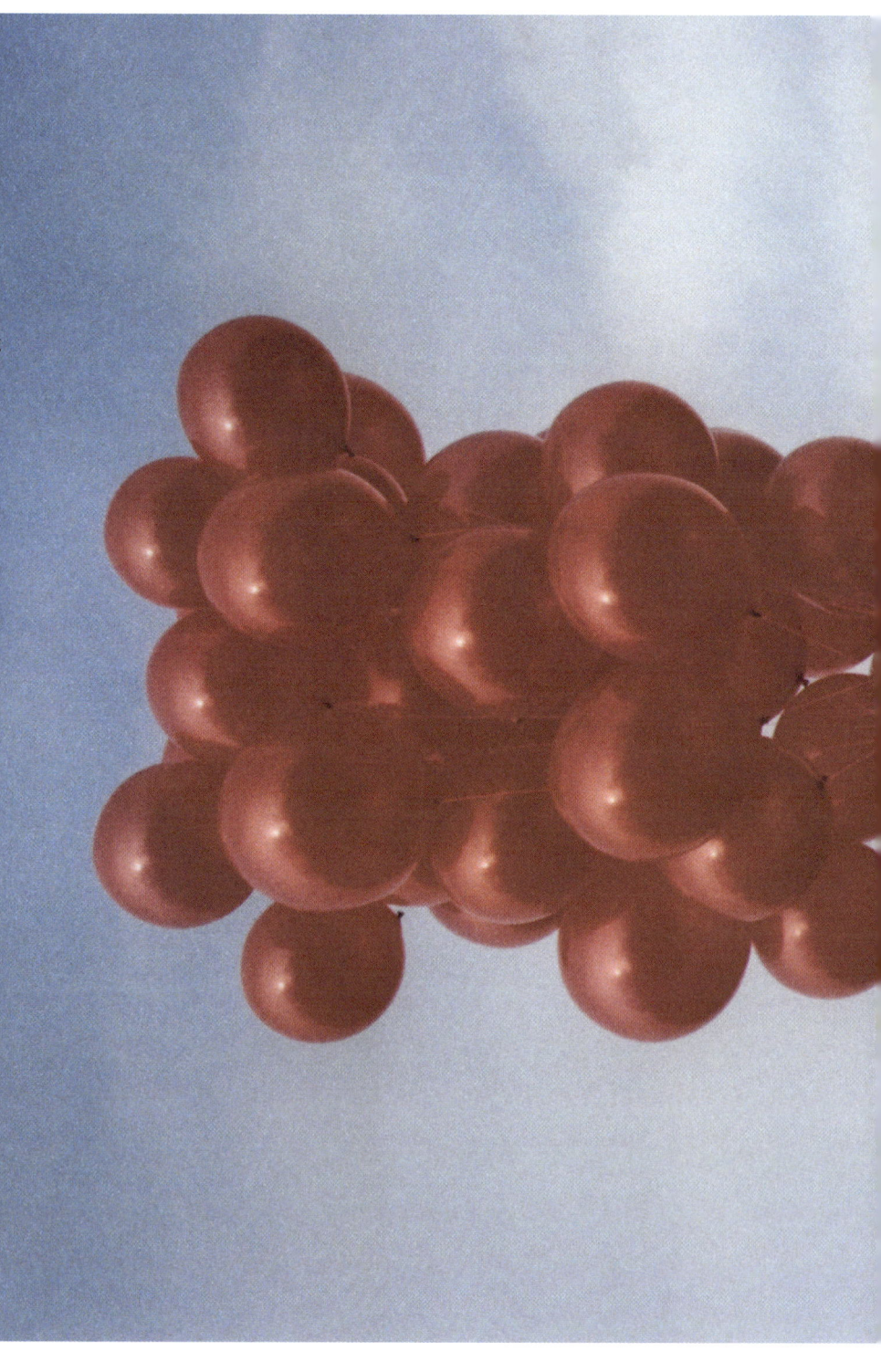

FUTURE VISIONING

Vision Concept

RICARDO MEJIA SARMIENTO & LIANNE SIMONSE

↘ Demonstrate the vision with an artifact

To demonstrate a future vision, some companies create vision prototypes, such as concept cars and concept kitchens. These vision concepts are created to explore and discuss new strategic ideas for innovation. According to designers at Citroën, concept cars are like "laboratories for new ideas[15]", while designers at Volvo say that "concept cars function as a test bed for new ideas[15]". The concept prototypes bring a future vision to life and make it possible to experience a future vision in the present[21]. In roadmapping, vision concepts do more than showcasing the visions of the strategic directions– they communicate the new values, test them by user interactions with the vision concept and support the strategic decision-making on the allocation of resources for future design innovations[8].

Although the embodiment of a vision concept can be quite similar to a 1:1 prototype, the inherent visionary narrative that it demonstrates is different from underlying build to test naratives of ordinary prototypes. Parts of vision concepts might find their way into new products, but a vision concept is not intended to provide a model for a production prototype – "nor will it be sold[15]" stated Philips designers explicitly in their write-up describing their concept kitchen. Vision concepts are therefore different from product development concepts, which unlock a problem and define the design challenge in a new product development project. Vision concepts are also different from emerging concepts that draw on proof-of principles of technologies in demonstration prototypes often found in pre-development projects. One step beyond these commonly-known prototypes, vision concepts have a strategic purpose and are specifically created to support a company's strategic decision-making on future innovation directions[15]. In-company, the vision concept influences and supports decision making on investments of additional resources into further research and design[21].

↘ Explore the future with plausible stories

A vision concept is a real, working artifact that demonstrates the plausibility of a future vision[22]. The intention of a vision concept is to explore the future potential and provide answers to research questions about future use, future systems integration and the social experiences

FUTURE VISIONING

A VISION CONCEPT is a publicly shared concept prototype to explore the future with plausible stories.

implied by the object[15]. A fully-functional concept car, for instance the F015 (Figure 3.1) serves the purpose of exploring the future with a broad audience of future users, media opinion leaders, competitors and in-company designers and engineers. Furthermore, after positive press feedback, it is easier to decide on further investment in immersive experience research[15].

↘ Vision sharing

A tangible vision concept encourages the creation and sharing of a clear vision[22]. As designers at Mercedes-Benz see it, concept cars are a way to have a "dialogue with customers[15]." Vision concepts are typically created for sharing future possibilities in a public context whereas other prototypes are built in secret and protected by non-disclosure agreements. Vision concepts are made public to showcase the organisation's strategic direction in innovation, to communicate

Figure 3.1
Vision Concept of Mercedes-Benz F 015 Luxury in Motion during its stay in the main square of Linz.

cc Florian Voggeneder photography.

the new values and test them by user interactions with the vision concept. The vision concept artifacts foster interaction to share and discuss the likelihood of use stories.

A concept car tells the story of a carmaker's future vision and innovation strategy. To illustrate the story, the vision concept is often featured in a short video created to capture the experience of interacting with it. For example, the Mercedes-Benz F015 research car (see figure 3.1) comes with the brand message "Luxury in Motion: an innovative perspective into the future of mobility[15]." The video Mercedes-Benz produced, features a handsome businessman and a sleekly futuristic self-driving car, to showcase the F015's immersive luxury user experience. The autonomous vehicle waits in standby mode until the businessman summons it, after which it chauffeurs him along the highway while he sits back and relaxes in a lounge chair turned away from the tiny steering wheel. From the interior – the "digital living space" – he interacts seamlessly with the outside world by novel web-services, using touchscreens installed all over the concept car's side panels[15].

Mercedes-Benz used this connected concept car and its video presentation to publicly announce its innovation strategy for autonomous-driving cars. The F015 compellingly demonstrates the company's vision of autonomous mobility innovation and luxury digital connectivity. It also indicates the strategic direction of their investment decisions. To further immerse the public in its vision, and test the market potential of the new user values, the fully-operational F015 Luxury in Motion was shown at CES exhibition in Las Vegas. Visitors were invited to test 'ride' the connected car. Eventually, Mercedes-Benz uses concept cars to interact with its customers more directly and influence and "set industry trends, preparing the way for market adoption of novel vehicle concepts[15]."

In other words, concept cars drive innovation in the automotive industry by providing inspiration to the market and exploring the customer interest for the new values. In a way, vision concepts can be used to assess whether the value drivers of the future vision have an extraordinary quality of becoming a self-fulfilling prophecy by building positive hype and visualising future use[15].

Virtual artifacts

Besides clay models and 1:1 prototypes, new technologies like augmented reality and 3D modeling also allow designers and innovators to explore the future of strategic innovation[15]. Virtual techniques are becoming increasingly useful ways to engage user interaction and explore the potential of new values and technologies to foster the development of novel platforms, products and services.

Creative Lead

↘ Taking the lead in visioning

Future visioning often implies leadership, especially when it comes to shifting the vision from the stages of imagination and creation, to the realisation. Studies have found that in environments characterised by uncertainty, successful leaders are those who sustain an ongoing vision[23]. And not just that – an effective leader must also strive to strengthen the bond and attunement among group members[24]. If only one person takes sole responsibility for future visioning directing the course of change, it is not just creativity that may get lost – the team's commitment to the vision may dwindle if participants do not take personal pride in pursuing the vision[24].

Research on leadership practice has shown that the abilities to champion and secure a vision are harder to master than those of to initiate and direct a vision[23]. For a vision to become a reality, there is a need for leaders who master these extraordinary abilities – leaders who are capable of communicating the vision and making sure others "buy into" it, and who can transform a vision into the sustained effort and ongoing cohesion required to make it a reality[24].

Every single person on the design roadmapping team can potentially exert an influence on the future direction of an innovation strategy. At certain times, creative individuals, who are skilled at imagining and visualising desired states, can take on a prominent role. At other times, those who are more in tune with users and their unmet desires can be the ones to take the lead. And then, individuals who are strong programmers and can make sense out of ideation sessions can be the ones to inspire others to transform their ideas into realities. In future visioning the designer can take the lead and catalyse the turn toward the future direction by, for instance, sketching scenarios of future use, identifying value drivers and visualising the vision concept.

↘ Turning the vision into reality

Research comparing the success or failure of new products on having a vision in advance has shown that radical innovations with a clear vision and a degree of flexibility built into project plans were the most successful[25]. A clear vision gives innovators a well-articulated, easy to understand target – a very specific goal that provides direction to everyone in the organisation. For example, the vision in the US space program, was to "put a man on the moon and return him safely to earth by the end of the decade." This vision easily helps others to create a clear image of what the innovation is

89   FUTURE VISIONING

Taking the CREATIVE LEAD in future visioning involves imagining, sharing, championing and securing.

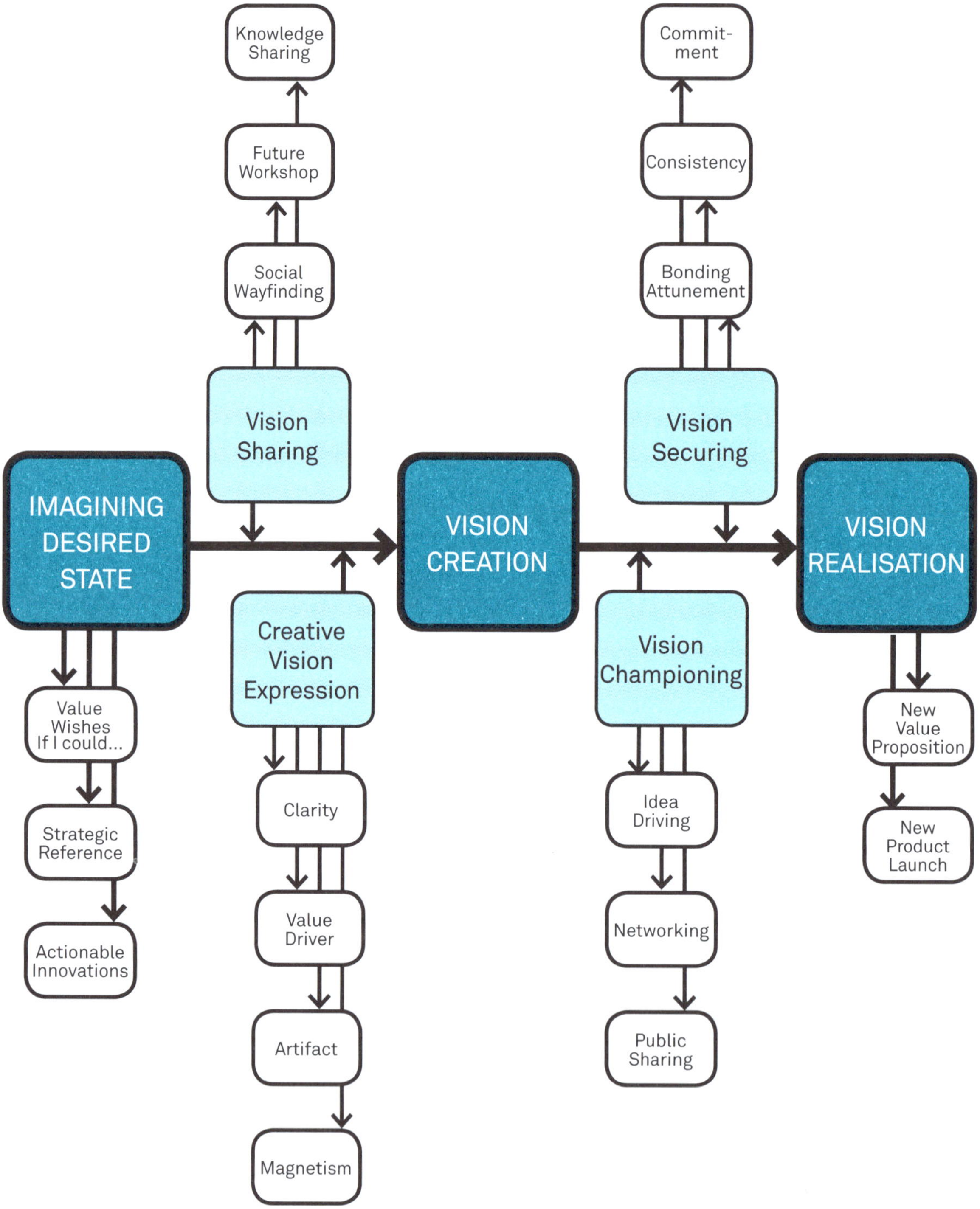
Knowledge Sharing
Future Workshop
Social Wayfinding
Vision Sharing
Commit-ment
Consistency
Bonding Attunement
Vision Securing
IMAGINING DESIRED STATE
VISION CREATION
VISION REALISATION
Creative Vision Expression
Vision Championing
Value Wishes If I could...
Strategic Reference
Actionable Innovations
Clarity
Value Driver
Artifact
Magnetism
Idea Driving
Networking
Public Sharing
New Value Proposition
New Product Launch

←
Figure 3.2
Future Visioning
an Integrative framework

cc Simonse & Hultink, 2017[8].

trying to achieve[25]. Establishing such clarity is essential in creating a successful vision. Without it, others might not support the vision – because they don't know what they are supporting – nor is the vision likely to be stable and enduring[24]. Figure 3.2 presents a framework of future visioning[8]. It gives a recap on the activities that creating a future vision requires. In the previous section we have discussed that creatively expressing a desired future, imagining a desired state of the future, sharing a vision and turning the vision into reality are key activities in future visioning.

In roadmapping, the challenge is to formulate a future vision, one that clearly points to new values. Often on roadmapping teams, the creative lead is challenged by overcoming the tensions between market pull (marketing professionals emphasising new market creation and new value drivers) and technology push (inventors and engineers pursuing the long-term advancement of technology). 'Building bridges' in the team communication and beyond requires creative ways of facilitating interaction with the roadmap.

↘ Championing the vision

Creative people, who take charge of communicating the vision, are those individuals who embody the key values and ideas contained in the vision and 'walk the talk'. They use stories, metaphors and analogies to paint a vivid picture of what the vision will accomplish – these so-called "champions" achieve high impact on innovation performance[16]. Ikea's video on the "Concept Kitchen 2025" has such a creative lead, who expressed the purpose of the concept kitchen: "to inspire ourselves and inspire people around us by communicating the behaviours of the future, and what the kitchen will look like in 2025[15]." Vision champions are particularly adept at propelling vision realisation: (S)he has mastered the skills of idea-driving communications and networking. For both skills the context of communication and networking can vary from communication in the roadmapping team to the communication on an expo. In roadmapping practice, the vision champions in roadmapping encounters at least three contexts of sharing the future vision:

→ TEAM SHARING where the champion delineates the future vision alongside innovation team players, and motivates researchers and designers to find answers to exploratory research questions;
→ IN-COMPANY SHARING where the champion inspires the creation and exchange of ideas with and among internal innovation professionals and external strategic partners; and lead customers.
→ PUBLIC SHARING where the champion shares a strategic intent of the future vision with an audience of consumers,

opinion leaders and potential partners (and competitors by for instance showcasing the vision concept (a concept car, concept kitchen etc.) or another type of artifact[15].

According to Harvard and MIT Professor Deborah Ancona, good leaders are those who are able to frame visions along certain key value dimensions. She encourages leaders "to practice creating and communicating a vision in many arenas, and enable co-workers by pointing out that they have the skills and capabilities needed to realise the vision[26]". In this way you can effectively nurture vision creation into realisation.

↘   Securing commitment

Attaining support for the future vision means securing the commitment of people throughout the organisation for what the overall vision is trying to achieve. When everyone is more than willing to pitch in and help realise a vision and will do whatever it takes to achieve that goal, the vision is securely supported.

Securing the vision includes the communication aimed at reducing individuals' natural resistance when they perceive that change is being imposed upon them[24]. Two important encouragements in these situations are: to 'suspend instant disbelief' and to concentrate on the long term aspect of the future vision for which the roadmap supports achieving stability, by providing consistency of the vision over time[8]. To secure commitment – the roadmapping team must operate as a well-oiled machine. This is possible when team members transcend traditional function requirements, and cultivate a greater sense of community, trust, respect and shared values in the interest of getting the job done.

Overall, the integrative framework in figure 3.2 outlines the leadership qualities that future visioning requires. In addition to imagining and creatively expressing a desired future, the key skills to master for turning the vision into reality are those of championing the vision and securing commitment[8].

FUTURE VISIONING

↑
Wander the myriad pathways! Skydive, metaphor for vision securing.

cc Treklocations.

FERRARI CASE

Spaceship Vision
FLAVIO MANZONI & ENRICO LEONARDO FAGONE

Flavio Manzoni is passionate: not just about his work, but also about futuristic, science-fiction inspired design. One of his signature projects was a study for a Ferrari-inspired spaceship, a creative exercise for his team in looking to the future for inspiration. While maintaining respect for Ferrari's heritage, he takes inspiration from the future, instead of building on the 'retro' style legacy or existing constraints[27].

In 2015, Manzoni unveiled his concept for a super-futuristic spaceship. He designed the Ferrari spaceship as a carrier of his vision to design cars and products that reflect progress and evolution.

> "I tried to imagine something that can fly into the future, since there will be less and less space available on the ground, and I focused on creating a little craft that's different from my childhood dream[32]."

In his opinion, the artefacts created for it, such as the drawing and the 3D rendered video in figure 3.3 and 3.4, encourage our trust regarding the future, like when we were young, and had a strong sense of trust in the future. He vividly remembers the feeling like the future was something to look forward to. As a young boy, he imagined that cars would be totally different from what they are now, "I expected to see flying cars by 2000[30]."

The Ferrari spaceship offers an alternative to the tradition of defining a new car design through standard modules such as the grill at the front and the bumper at the back, and studying the fronts and backs of previous models to derive inspiration. He wanted to create something new instead- something inspired by the future[27]. For him it's simply not enough to create a new version of the front and the back of a car. His ambition is to achieve a true union of technology and aesthetics that has been guided by the future. He does not like to start designing from existing constraints, which are always a huge obstacle. In his view it depends on how you face them, he does not think of them as a problem, because then the end result will be compromised. He considers it

---

Figure 3.3
Sketch Spaceship vision

© Flavio Manzoni,
FERRARI DESIGN, SVP.

Figure 3.4
3D artifact of Spaceship vision

© Concept Flavio Manzoni,
3D model Guillaume Vasseur
Rendering and post production
Billy Galliano,
FERRARI DESIGN TEAM.

95 FUTURE VISIONING

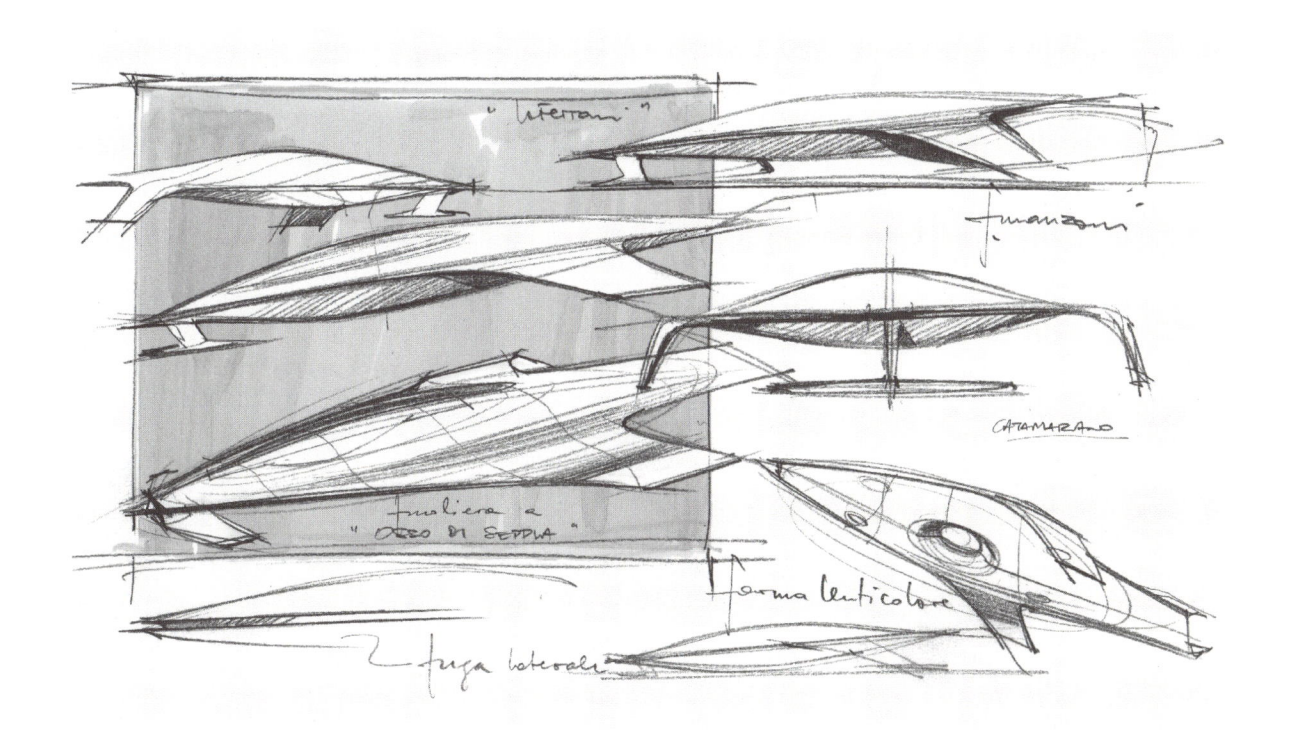

something that forces us to do something different, then that opens up a massive opportunity to say something new, to transform and evolve the design of the car.

↘ Creative lead

According to Flavio, the 'retro' design trend has become over-popular, especially in the last 20 years. Every car company is making retro-styled products, whose inspiration comes from the past instead of the future. For him, this practice does not fit with his personal drive to design the future. As Senior Vice President of Ferrari Design, and alongside the Ferrari Styling Center, he seized the opportunity, to try and set a new standard -to raise the bar -, creating an iconic, modern language.

He came into Ferrari, a brand known for its history, its heritage – Enzo's heritage – and its timeless design and gave a radical new twist to the meaning and creation of 'timeless' cars. "Sometimes it's not so easy to explain why we feel that a particular design solution is the correct one. Normally I follow this principle: if the form follows the function, and our work is innovative, then we are safe. The problem is when you try to make something stylish, when you have a preconceived idea of the form, then it becomes an ornamental treatment[30]." In his view this is never timeless. When, instead, there is a clear connection, a clear match between the form and the substance, then it is different and the result can be timeless. He wants the team to imagine when they work on a new Ferrari,

"Please, guys, think about how this car will be perceived in the next fifty years[30]."

↘ Vision concept artifact

Giving form to a future visioning project is quite a difficult task, because although the ufo-idea that he had in mind when they started the project was quite clear, it wasn't clear how they could squeeze in all the requirements, all that technical complexity. They found inspiration in this quote from the architect Oscar Niemeyer: "It must be extraordinary not only in terms of performance, but also extraordinary in terms of beauty. I recall you saying that the voids are as important as the solids [31]."

For this project they found a new way of designing by modeling the airflow, because once we knew what the behaviour of the air would be, then we could imagine the form. Maybe in a fifty years we'll finally have our personalized spaceships. By developing the FXX K and

↓
Figure 3.5
Ferrari 812 Superfast, model presented to the public of the International Car Show in Geneva in March 2017.

© FERRARI MEDIA

DESIGN by Flavio Manzoni and team. The design of the object conciliates efficiently the extreme car performance ever achieved with the harmony and proportions of a 'Granturismo'.

other concepts. Flavio aims to build a portfolio of cars that look like spaceships, inside and out.

"We start very visionary, and then we have to domesticate that vision. But only a little, enough to ensure a Ferrari is recognisable without the emblem[29]."

The propulsion method remains undetermined, and yet this concept car is only years away from becoming a reality.

↘ The use of artifacts in visioning: intuition and 'heuristic' approach

The final shape of a car is the consequence of a complex design process based on integration. Flavio Manzoni has always been a proponent of a rigorous approach in which the designer has to combine the essential technical and performance requirements. The latest Ferrari models can be considered the result of the synergy with the technicians and specialists of aerodynamics as well as the consequence of a proper conceptual approach and working methodology. "We can identify a match -Flavio Manzoni says- which expands the scope of action and reflection of the designer in the broader context of visual culture, art and science, architecture and music".

In LaFerrari, for example, surfaces define a continuous harmony that recalls the research into abstract forms and the '3D surfaces' by the artist Anish Kapoor and the studies on the 'fourth dimension' of the mathematician Bernhard Riemann. Similarly, the project J50 can be understood and read as fruit of a creative process close to the Jazz music compositional criteria and based on dialectic. Even in the case of the 812 Superfast, the new Ferrari model that has been presented to the public of the International Car Show in Geneva in March 2017, the design of the object conciliates efficiently the extreme car performance ever achieved with the harmony and proportions of a 'Granturismo' (see figure 3.5). "In a way I follow a 'heuristic' approach to the solution of the problems that don't follow a logical path. I rely a lot on intuition and for the transitional state of the circumstances, in order to generate new solutions[28]."

For Flavio Manzoni and his team, the creation of a new product is to be considered always a special moment, the search for a form that has to materialize its very essence:

> "It's a challenge with oneself, with the mind turned towards future scenarios. A collective-individual process, of initial expansion and final convergence. The process that leads from a blank sheet of paper to the finished car, is a magical occurrence that is renewed every time a new car is born, from the

first pencil strokes, with an ideal in mind and the 'creative fire' that pushes the hand to lend it a shape. When those traits come to life, my dream is that this ideal can permeate the project, crossing the entire process of development and arriving intact to characterize 'that' product, speaking a language, or a 'meta-language', able to communicate without words".

Flavio Manzoni's desire is such that this creation is self-explanatory: it expresses its philosophy, its values and its inherent innovative content through its own formal language. But above all that it communicates directly the emotion that has generated it.

"I learned to work in an interdisciplinary way, seizing the opportunities of those 'short circuits' that come from putting in connection fields, which differ from each other, from architecture to sculpture, from industrial design to music.

It's the so-called 'serendipity': creativity means for me the ability to see beyond relationships where they still don't exist, the rapid act of connecting elements that do not belong to the obvious[29]."

FLAVIO MANZONI is Senior Vice President, Head of Design at Ferrari. He graduated in architecture with a specialisation in industrial design, and began his career in 1993, working for Fiat, Lancia, Volkswagen, and Seat, before joining Ferrari in 2010. Under his creative leadership, Ferrari was awarded with the prestigious "Compasso d'Oro" for the F12berlinetta in 2014 and the FXX K in 2016, which also received the "Red Dot: Best of the Best" in 2015. One of his masterpieces is considered the LaFerrari, the first Ferrari hybrid as well as its most powerful production car to date.

DESIGN ROADMAPPING

LAB ↗

Expression on a desirable future

MATERIALS NEEDED:
- → sticky notes
- → big sheets of paper
- → white board or wall

1 Form a subgroup of 3-5 design and innovation professionals from the roadmapping team.

2 Start the conversation about a future vision by telling each other which value desires and wishes you're interested in, and which you consider to directly affect your organisation. One by one you'll go around the room, and capture your perceptions, ideas, sketches and stories on sticky notes. It's critical to pay close attention to your team mate's stories, learnings, and hunches.

3 Be open to each other's reactions and interpretations. Encourage the association of ideas on how users and organisations are, or could be responding to these value desires and wishes. Generate a bunch of ideas on strategic value opportunities in the future. These may include but are not limited to new user experiences, new technology interactions or new service desires. If there is an idea that does not resonate, drop it and move on to the next.

4 Put those value opportunities that resonate in the roadmapping team on the big sheets of paper so that you can start formulating the vision statements by its strategic value drivers. Gather your group around the sheet with sticky notes. Move the most compelling, common, and inspiring values together and sort them into categories.

5 Look for patterns and relationships between your categories and move the sticky notes around as you continue grouping. The goal is to identify key strategic themes - value drivers - and then to translate them into the vision statements. Arrange and rearrange the sticky notes, discuss, debate, and talk through what's emerging. Don't stop until everyone is satisfied that the clusters represent rich value drivers.

6 Take the value drivers that you identified and put them up on a wall or board. Now, take three (max. five) of the value drivers and rephrase it as a short statement and sketch an image that illustrates it.
- People don't need to be a great artists to create great images on the futute.

This lab provides guidelines on the team activity of future imaging and how to converge that into the formulation of a future vision expression. It proposes to organise a creative conversation with the result objective of an agreed upon future vision expression. In a shifting and uncertain world, a well-defined desirable future is often expressed in three to five strategic value drivers.

At the same time, you are working on making your organisation as actively adaptive as possible, in relation to achieving those values and in relation to changes that will occur in the external environment. An agile organisation has the ability to align their actions to value drivers, as well as modify, drop or add a particular value driver as the environment changes by recurring creative conversations on the desired future.

Prof. dr. GLORIA BARCZAK on the subject of Visionary Leadership and Roadmapping

SCIENCE INTERVIEW

LS In roadmapping, the creation of a future vision is providing the destination of the roadmap. You identified that shared visioning is important in organisational change. To what extent do you consider this also relevant to innovation strategy?

GB Well, so I think two things, when organisations want to move, from doing only incremental types of innovations to doing some breakthrough innovation, then clearly they need a change in culture. But before they can do that they need to establish a vision of what they want to achieve and why this vision is needed; and then create the roadmap of how to get there. Think about P&G with their open innovation initiatives; at the start of the 2000's A. G. Lafley was CEO and says we need to do open innovation, and we need 50% of our innovations to come from outside the organisation. You are not going to get breakthrough innovation, if you are not setting an actionable vision. You are also not going to have an organisation that delivers continuous innovation unless a leader sets some kind of a vision and discusses its implications. The leader needs to tell people: 'this is what we want to do, explaining what we are going to do, why we are doing it, and why it is important for the organisation'. Without communication and explanation about the vision and its implications, people won't get it.

> LS Are there parallels between a corporate vision and a future vision for a long term innovation strategy?

GB I think there are parallels. Organisations, if they are really thinking about how are we going to survive for the next 5-7 years, then you have to have an understanding of what is going on in the environment. In my experience, often there is a lot going on, and nobody knows exactly how things are building up. I think it makes sense that if you want to continue to survive as an organisation, you have to think about the future. Although nobody knows for sure, but you can see signs of changes and trends, read those signs, and collectively you have to think about what are the implications. That is why people use scenarios, even though that scenario may not materialize exactly. It is important to think about the implications, "what is changing, how will it impact our stakeholders and our company, and what do we have to start doing now to address these changes and trends?" So that is where roadmapping comes in. If we are thinking that some change is really going to take effect several years from now, we have to consider what do we have to do now and how do we map that out, to get to that point.

> LS Your research on visionary leadership showed the importance of bonding and attunement in securing a vision. Do you consider that visionary leadership stems from a personal ability, a charisma, of leaders or a way of acting?

GB I do not think that everybody has visionary skills; some of it is innate. When you compare leaders, people like Elon Musk and Steve Jobs, they are really visionary leaders. Jobs was able to see things that people could not see. But there are not a lot of people like that. Most leaders set goals 3-5 years ahead but that is different than being a visionary leader. Those who are visionary, look ahead all the time. They are very future oriented, and that is part of who they are. Therefore, I don't think you can teach people to be visionary. You have it or you don't. Steve Jobs had it, Elon

Musk has it, it is innate; it is a personality characteristic. This implies that if you want visionary leadership, you have to find people who have that ability. I'm not sure how you assess visionary leadership or measure it. My sense is that, when you compare visionary leaders to non-visionary leaders, the visionary ones, can see way far in the future. Not 5 years ahead but 20 or more; with a much longer term perspective. When I think about myself, when I took over the journal, I could see things that needed to be done for the next three years, but I don't know what is going to happen 15 years from now on. There is a difference between having goals and strategies and then having a vision.

> LS So for designers, who are thinking of taking the creative lead in roadmapping – and do not have to look that far into the future - What would you advise?

GB You can share a vision, by sharing your thoughts and reasons why, and hoping that people buy into that. To create commitment you have got to involve other leaders. Particularly, in larger organisations, you need leaders who can implement and champion the vision; leaders who can communicate the vision and persuade and motivate their people. Sharing the vision involves shared leadership for the vision. You have the visionay leader, yet you also need to have people with situational leadership, those leaders that represent the different levels of the organisation

> LS And what about creative leadership. Recently there is a lot of discussion about creative leadership and transformative leaders?

GB Well leaders sometimes have to be creative in getting things done. But a creative leadership kind of implies that some leaders are creative, and others are not. This goes back to the myth that only certain people are creative and others are not. I am convinced that all people are creative, although in different ways, and some are able to tap into their creativity more than others. The job of a leader is to leave room for creativity. The autocratic person who might be very visionary, often does not leave room for creativity. It is his vision, you are not allowed to be creative in terms of where we could go or how to get there. This type of leader is not interested in getting commitment to their vision; they are only interested in having it implemented.

LS My last question: what do you recommend design roadmappers in terms of leadership?

GB As we have been talking about, the first thing would be to create a shared vision, and really organise this collectively. You have to involve people from all levels within the organisation. Obviously the people at the top can see things and know things, but certainly at all different levels people see things, often things that the top has no clue about. Sometimes for the people at the bottom it is difficult to see the future, because they have to deal with the day-to-day management. However, you can't have commitment to a vision from solely the senior management. A real trap is that their 20 years' views can get so disconnected from the people at the bottom.

Second, really think about what that future might look like in the next 5 years (instead of the full 20 years). Think in increments; here is where we are, here is what the future is likely to look like in 5 years. Then, consider what innovations and changes are needed to get to that future. Which innovations can you map towards that vision. What innovations do you have to create or buy, in order to come to that future vision.

Third, we have talked about that notion of commitment and buy in, that really has to be there in order to succeed.

GLORIA BARCZAK is Professor of Marketing in the D'Amore-McKim school of business at Northeastern University in Boston. She is the Editor of the Journal of Product Innovation Management, a senior advisor to Creativity and Innovation Management, and a member of the editorial board of IEEE Transactions on Engineering Management. She has published over 35 articles, one book, and several book chapters/edited proceedings. Professor Barczak is a long-time member of the Product Development & Management Association (PDMA).

↑
Self portrait of Tracy Caldwell Dyson in the Cupola module of the International Space Station observing the Earth below during Expedition 24.

cc Tracy Caldwell-Dyson, NASA.

IN SUM

Future visioning is about imagining desired values that are actionable and within reach of the participating innovation professionals. In this chapter, we have provided answers to the question of how to design a future vision. It all starts with determining the properties of the future vision. The three core elements of a roadmap's future vision are:
- → CLARITY in the explicitly expressed desired end state;
- → VALUE DRIVERS that capture the key compelling benefits of value wishes;
- → ARTIFACT that materialise the imagined value wishes;
- → MAGNETISM that energises others to aspire to its creation.

Next comes creatively expressing the vision in an artifact. We presented ultimate artifacts that we term 'vision concepts', which are nearly fully-functional, futuristic prototypes (such as the connected concept car and the concept kitchen). Then we outlined the abilities that define the role of creative lead for future visioning and highlighted some leadership qualities the job requires. Important skills for the creative lead to master are: IMAGINING, through dreaming and creative expression; SHARING, by establishing bonds and maintaining attunement; CHAMPIONING, by driving ideation, implementation and networking; and SECURING comitment until the vision has become a reality. Flavio Manzini's story inspires us to look ahead to the future, take a leap into actionable dreams and forget about the past and its retro legacy.

1. Slaughter, R.A. (1993). Futures concepts. Futures 25 (3): 289-314.
2. Polak, F.L. (1961). The Image of the Future. Enlightening the Past, Orientating the Present, Forecasting the Future. Leiden: Sijthoff.
3. Schwartz, P. (1996). The art of the long view. New York: Doubleday.
4. Reid. S. (2015). Vision and Radical Innovation: A Typology. In: Adoption of Innovation. Switzerland: Springer International.
5. Fiegenbaum, A., Hart, S. & Schendel, D. (1996). Strategic reference point theory. Strategic management journal, 17(3), 219-235.
6. Cooper, R. & Evans, M. (2006). Breaking from tradition: Market research, consumer needs, and design futures. Design Management Review, 17(1), 68-74.
7. Ziegler, W. (1991). Envisioning the future. Futures, 23(5), 516-527.
8. Simonse, L.W. L. & Hultink, E.J. (2017). Future visioning for innovation in the organisation: an integrative framework. 33nd EGOS: European Group for Organisation Studies Colloqium-SWG 42, Copenhagen, Danmark, 6-8 July 2017.
9. Van der Helm, R. (2009). The vision phenomenon: Towards a theoretical underpinning of visions of the future and the process of envisioning. Futures, 41(2), 96-104.
10. Turing, A. M. (1950). Computing machinery and intelligence. Mind, 59(236), 433-460.
11. Groenveld, P. (1997). Roadmapping integrates business and technology. Research-Technology Management, 40(5), 48-55.
12. Dunne, D. D. & Dougherty, D. (2016). Abductive reasoning: How Innovators navigate in the labyrinth of complex product innovation. Organization Studies, 37(2): 131-159.
13. Shipley, R. (2002). Visioning in planning: is the practice based on sound theory? Environment and Planning, 34(1), 7-22.
14. Heinonen, S. & Hiltunen, E. (2012). Creative foresight space and the futures window: Using visual weak signals to enhance anticipation and innovation, Futures, 44(3): 248-256.
15. Mejia Sarmiento, J.R. & Simonse, L.W.L. (2015). Design of vision concepts to explore the future: Nature, context and design techniques. In: Proceeding of 5th CIM Community Meeting, Enschede, The Netherlands, 1-2 September 2015.
16. Reid, S.E., Roberts, D. & Moore, K. (2014). Technology Vision for radical innovation and its impact on early success. Journal of Product Innovation Management 32(4), 593–609.
17. Buijs, J.A. (2012). Projecta's, a way to demonstrate future technological and cultural options. Creativity & Innovation Management, 21(2), 139-154.
18. Hekkert, P. & Dijk, M. van (2011). ViP-Vision in Design. Amsterdam: BIS Publishers.
19. El-Namaki, M.S.S. (1992). Creating a corporate vision. Long Range Planning, 25(6), 25-29.
20. Kantabutra, S. & Avery, G.C. (2010). The power of vision: Statements that resonate. Journal of Business Strategy, 31(1), 37-45.
21. Keinonen, T.K. & Takala, R. (2010). Product concept design: A review of the conceptual design of products in industry. Finland: Springer Science & Business Media.
22. Raford, N. (2012). From design fiction to experiential futures. Houston, TX: eBook by Association of Professional Futurists.
23. Weick, K.E. & Sutcliff, K. (2001). Managing the unexpected: Assuring high performance in an age of complexity. San Francisco: Jossey Bass.
24. Barczak, G., Smith, C. & Wilemon, D. (1987). Managing large-scale organizational change. Organizational Dynamics, 16(1), 23-35.
25. Tessarolo, P. (2007). An empirical investigation of the contextual effects of product vision. Journal of Product Innovation Management, 24(1), 69-82.
26. Ancona, D. (2005). Leadership in an age of uncertainty. Center for Business Research Brief, 6(1), 1-3.
27. Manzoni, F. (2015). Design for life key note speech. International Conference Engineering Design. Milan, Italy, 27-30 July 2015.
28. Manzoni, F. (2017). Keynote at the World Italian Design day conference. Berlin, Italian embassy, Germany, 1 March 2017.
29. Manzoni, F. (2016). The meta-language of shape. Lecture for Vivid Sydney, Australia, 30 May 2016.
30. Lagios, V. (2015). Design Farther interview with Flavio Manzoni. http://www.designfather.com accessed April 2017.
31. Niemeyer, O. & Burbridge, I.M. (2000). The curves of time: The memoirs of Oscar Niemeyer. London: Phaidon .
32. Gallina, E. (2015). Formtrends Cars, designs and the people behind it. http://www.formtrends.com accessed April 2017.

TECHNOLOGY SCOUTING & MO

109

TECHNOLOGY SCOUTING & MODULAR ARCHITECTURE

LAR ARCHITECTURE

HOW TO CREATE A TECHNOLOGY ROADMAP

Design roadmapping is rooted in technology roadmapping, which technology strategists invented to manage the growing complexity of technologies. In the early days of technology roadmapping, the objective for them was to make strategic choices on long-term technology research in alignment with the strategic decisions on new business development. Nowadays, technology sub-maps still provide the foundation for design roadmaps. Although in the digital age, not every organisation has an in-house technology research program, almost every organisation uses digital technology to aid product and service development. Therefore it remains vital for designers to align the value drivers of the user sub-map with smart choices about emerging technologies. Scouting for new technologies is one of the key activities to do this.

In the first chapter we introduced roadmaps that depict new design innovations along four layers of: user value, market, product and technology. The technology layer assists organisations to forecast and plan for new technology components and modules. Its structure is based on the partitioning of the system architecture.

This chapter takes a closer look at this strategic partitioning and delves deeper into ways companies uncover new sources of technology and can make technology evolution projections. The design of the technology roadmap is guided by (1) the mapping of technology evolution (2) the use of a modular system architecture and (3) source-based scouting. Paul Hilkens, the vice president at Océ-Technologies BV, provides us with a case study that highlights the integration of modular system architectures in technology roadmaps.

Mapping Technology Evolution

The evolution of new technologies – the progress from invention, to proof-of-principle, to prototype testing – can take years. To make optimal use of emerging technologies, and harness new user value opportunities, you can anticipate on these evolutions by mapping them. For instance, we mapped the evolution of LED (Light-Emitting-Diode) technology for Philips' wake-up light on the technology sub-roadmap. We placed special emphasis on the 2009 moment mapped at the future timeline when the proof of technology for "white LED-light" was to be expected – see figure 4.1 – because at that time when the roadmap was created, in 2005, there was only proof of technology for coloured LED lights. Before applying the white light LED technology into the wake-up light that (was introduced with incandescent light technology) we mapped the application of subsequently, energy-efficient halogen light technology, mapped on the year 2006, and durable compact fluorescent light technology, mapped on the year 2007.

The mapping of technology evolution has three main activities:
  (A)   Out-of the box scouting to discover new and promising technology applications that support the visionary direction;

TECHNOLOGY SCOUTING & MODULAR ARCHITECTURE

MAPPING TECHNOLOGY EVOLUTION by building logical chains of scouted technologies.

(B) In-depth scouting for plausible and highly promising application ideas and technology options that extent current product lines and also provide for new value propositions;
(C) Mapping the evolution of strategically interesting technologies for further internal development or partnership collaborations in design innovation.

Figure 4.2. shows the evolution of display size technology mapped on a timeline. Such evolutions are commonly mapped on the future timeline of a technology roadmap[1]. Yet, it is also important to bear in mind, that a chain of anticipated technology innovations is not always a straight line or extrapolation - it might involve a jump towards a new emerging technology innovation.

Out-of the box scouting

The out-of-the box scouting results in a status overview of emerging and upcoming technologies. Along the collected evidence of patents, demonstrators with proof of technology, or samples of new 'state-of-the art' technologies, new components from supplier companies are listed. Discoveries from technology research in progress can lead to promising applications in design innovations that the future vision may require. Be aware, that although it is important to detect technologies – in their

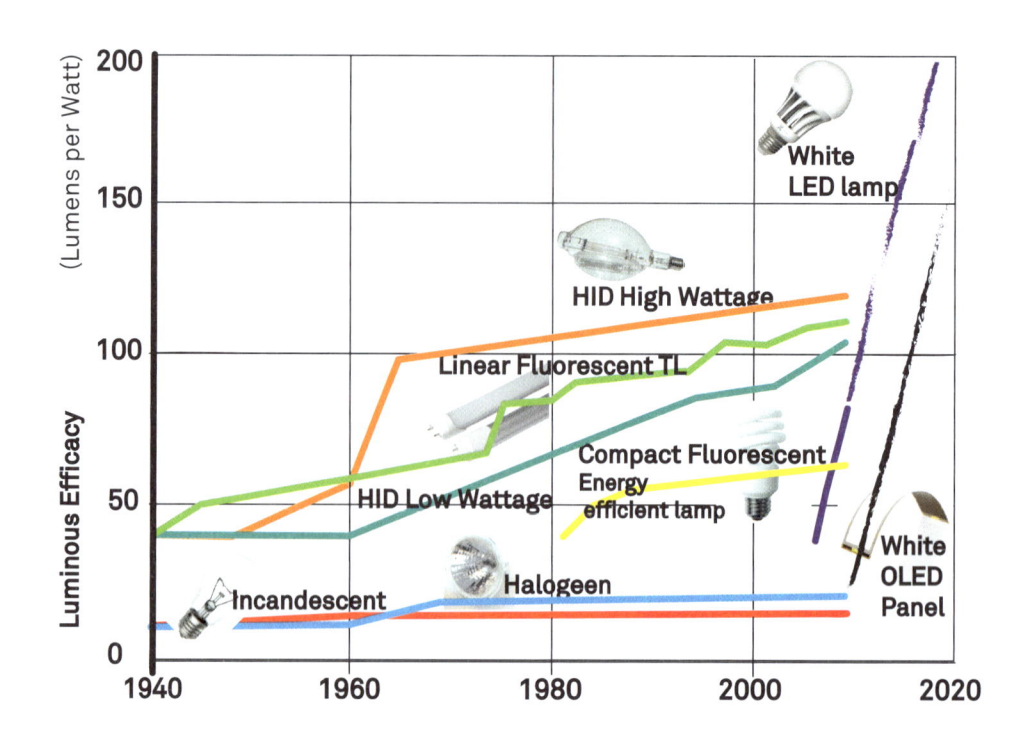

Figure 4.1.
Lighting technology evolution.

cc Martin Klaasen,
Lighting Design.

Figure 4.2.
Display size evolution.

cc Matti Matila.

broadest definition, too generic technologies as for instance artificial intelligence - AI - or 3D printing trends, are not the preferred outcome of technology scouting. Beyond identifying an upcoming technology, scouting also includes ideas of application of the technology. Take for instance the application idea of self-producing drone parts with 3D printing such as in the Drone parts by Fusion Imaging. Their application idea was to support the Phantom Drone consumers by offering them the digital files of unique parts they can 3D print at home for the do it yourself customisation and repair of their drones[5]. Another example is the Nest's application idea of AI in their smart thermostat concerns a particular sensor and a control software platform[6].

In-depth scouting

For the in-depth scouting, the current product is deconstructed into its modules and system architecture[1]. Then the supply market of each module can be investigated[2]. Besides this downstream scouting of technologies, the upstream scouting can gauge new inventions that might be patented. A third, in-between stream of technology scouting includes fringe markets and industries – technology applications in other industries that can serve as inspirational examples of ways to re-engineer, restyle, upgrade, downgrade and isolate new technologies.

Mapping the evolution of strategic technologies

After the scouting activities have generated substantial input, the activity of mapping the evolution firstly requires decisions regarding the strategic importance and urgency. Not all scouted technologies are mapped, only those that have user relevance and strategic fit for a new product or service

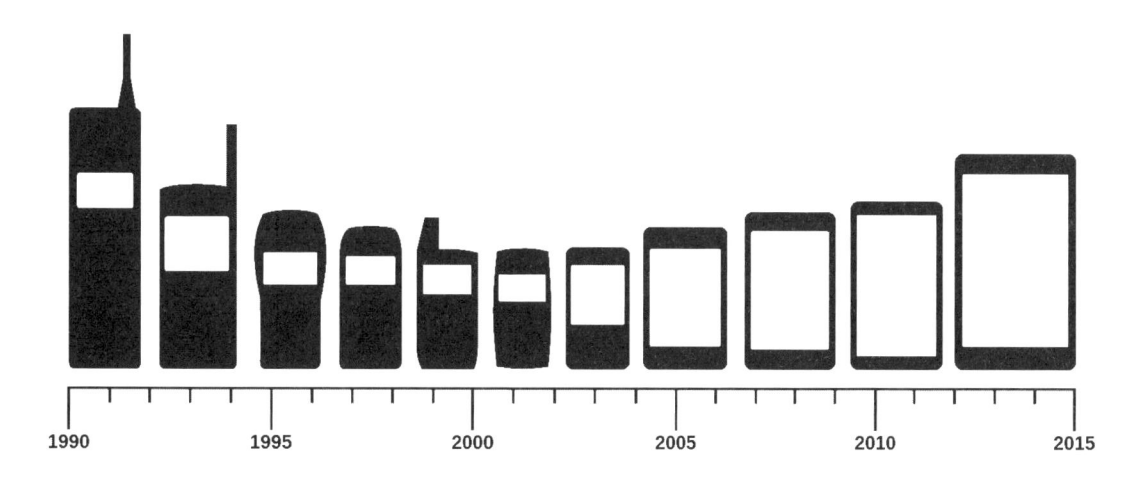

TECHNOLOGY ROADMAP
Lucent Technologies

→Figure 4.3

Technology Roadmap example.

cc Richard E. Albright, 2003. Bell Laboratories, Technologies Office, Lucent Technologies[4].

The technology roadmap, shows sequences of scouted technologies mapped on the modules of a particular cell phone business.
See for instance the evolutions mapped for the microprocessor with two successive models, 8523 and 8524, followed up by a single chip processor in its research phase. Or the user interface module that shows the evolution of keypads, softkeys towards the application of voice recognition technology.

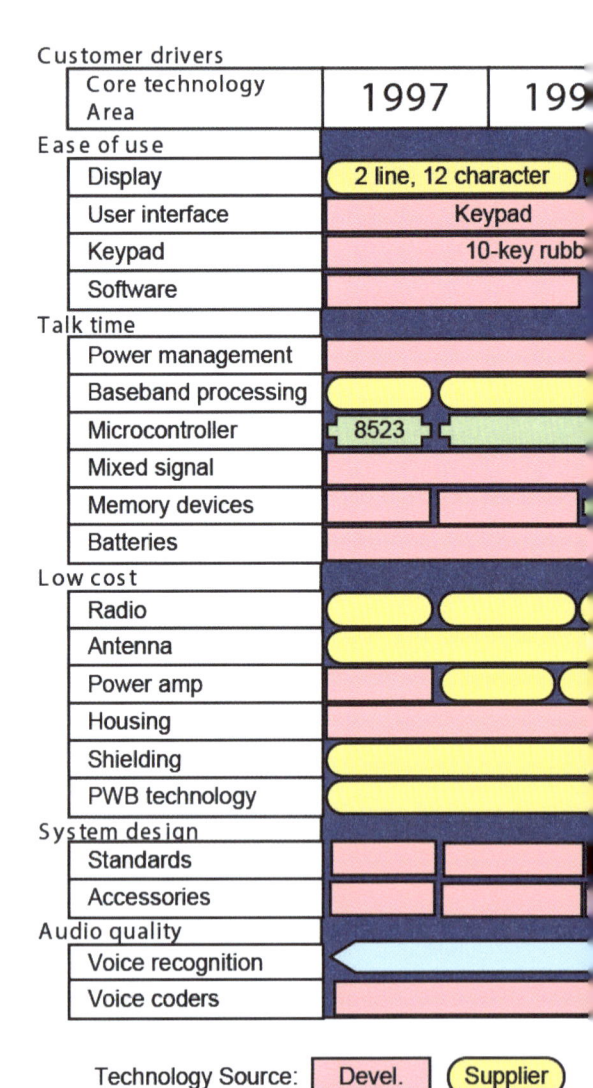

Customer drivers / Core technology Area	1997	199
Ease of use		
Display	2 line, 12 character	
User interface	Keypad	
Keypad	10-key rubb	
Software		
Talk time		
Power management		
Baseband processing		
Microcontroller	8523	
Mixed signal		
Memory devices		
Batteries		
Low cost		
Radio		
Antenna		
Power amp		
Housing		
Shielding		
PWB technology		
System design		
Standards		
Accessories		
Audio quality		
Voice recognition		
Voice coders		

Technology Source: Devel. | Supplier

TECHNOLOGY SCOUTING & MODULAR ARCHITECTURE

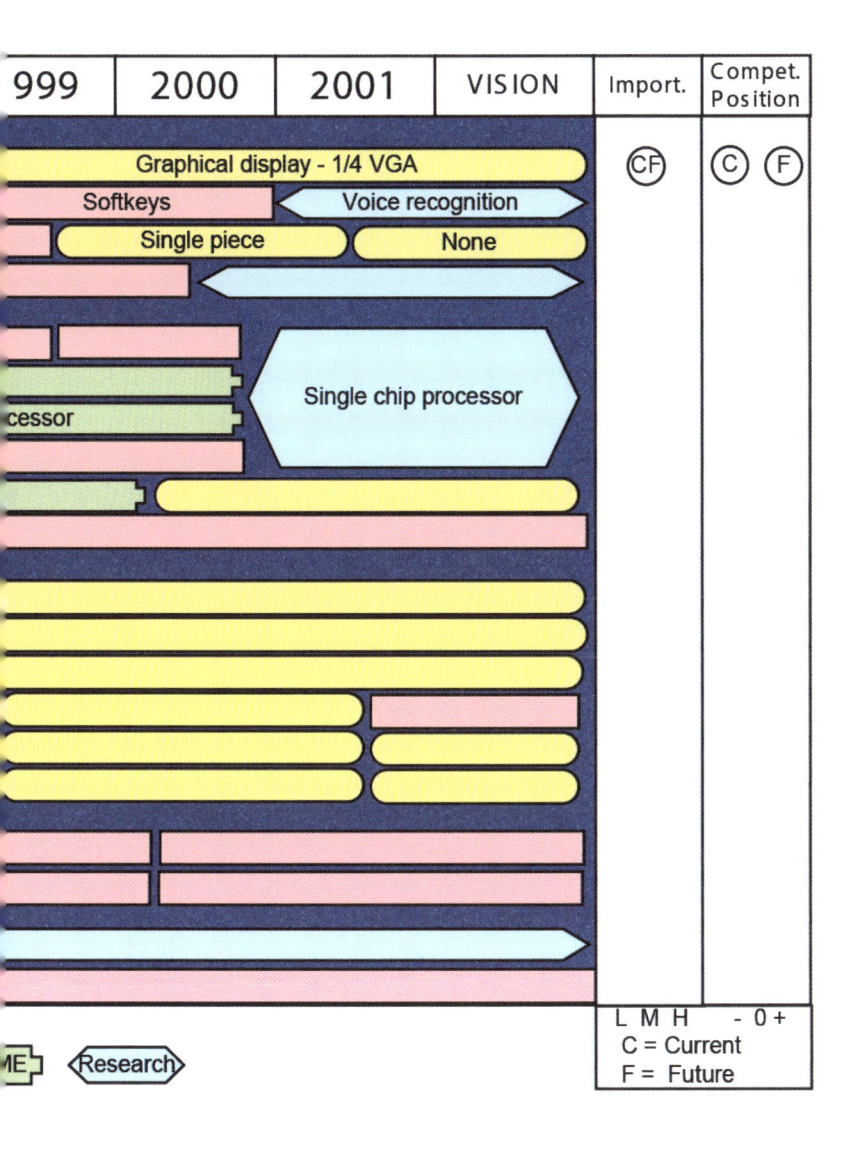

999	2000	2001	VISION	Import.	Compet. Position

Graphical display - 1/4 VGA
Softkeys
Voice recognition
Single piece
None

Single chip processor

L M H   - 0 +
C = Current
F = Future

Research

innovations. A decision grid can guide the strategic choice (see figure 6.5). For choosing the technology application options with the highest potential two criteria are used. 'Is the impact on user value real?' and 'Can we do it?'. Only high user impact and high strategic fit options are selected for mapping on the technology roadmap. For the others, the decision is either to drop it, or park it on a list for periodic review.

Second, the evolutions of only the strategic interesting technologies are mapped: those pertaining to distinctive components that – with further development– can contribute to future innovations and provide competitive advantage by creating new user value[3]. Such as for example, the user interface (UI-)module in the Lucent roadmap, shown in figure 4.3, maps the evolution of 'keypad' into 'softkeys' and after that the application of 'voice recognition'[4]. Depending on the decisions of the roadmapping team, an evolution is sometimes an extrapolation, sometimes a jump to a new breakthrough or disruptive technology and sometimes an evolution in technology performance. Roadmapping such technology evolutions thrives on gathering substantial input by out-of-the box and in-depth scouting.

Modular System Architecture – the radar

To effectively identify new technology options, you will find it advantageous to gain some preliminary technical know-how regarding the modular system architecture . Mapping technology evolutions of, for instance, personal computers (PCs) involves scouting for upcoming updates to modular components, such as the microprocessor chip, hard disks, memory chips, display, graphics card, speakers and so on. The modular system architecture makes it possible for roadmappers to map out innovations to a product line or a family of products and services with successive versions of upgraded products[2].

Technology scouts consider their system architecture as a kind of radar that provides a scope to detect and track changes in relevant technological areas, including at the fringe areas. In essence, the architecture defines the essential technical structure of the system; it specifies the partitioning of the overall functionalities of a system into specific functional components[3] . Modular architectures, characterised by a 'one-to-one' coupling of function and module, allow for agile design of a single module independently of the other modules[7]. The next versions of the new product replaces a module with the latest innovations in technologies . Modular architectures enable long-term flexibility of upgrading products with distinctive and desirable modules without the need to completely redesign the whole system[3,7]. The design of a modular architecture commonly requires a project effort of a system architects[2].

TECHNOLOGY SCOUTING & MODULAR ARCHITECTURE

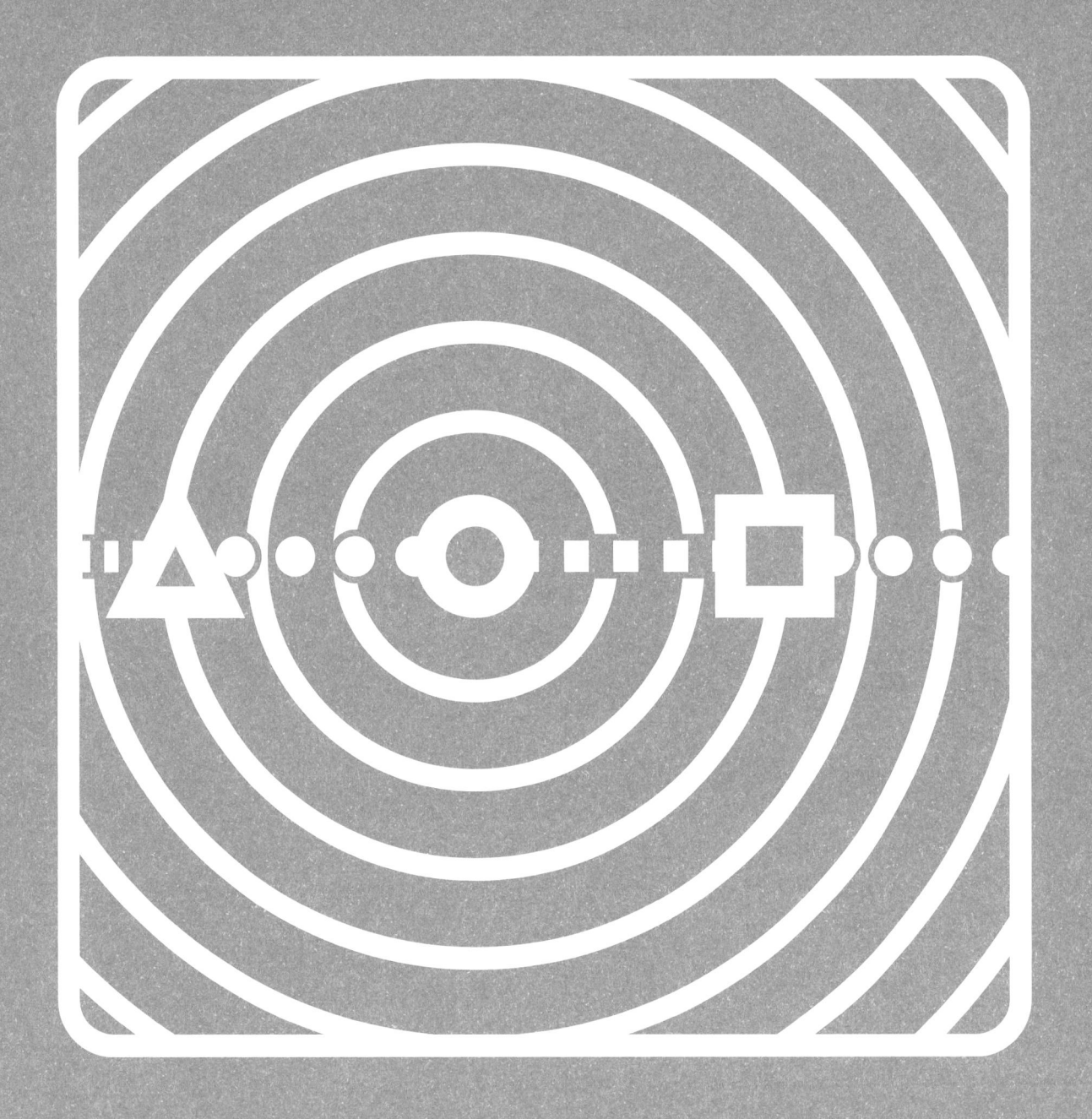

a MODULAR SYSTEM ARCHITECTURE specifies the partitioning of the functionalities of a system into specific modules, and their interfaces.

Teardown the system

→
Figure 4.4
FAIRPHONE 2 Teardown,

cc iFixit.
Published: November 18, 2015 on https://wwwifixitcom/Teardown/Fairphone+2+Teardown/52523.
Licensed under the open source Creative Commons.

The disassembled FAIRPHONE2 is used for the partitioning into a basic system architecture with:

(A) Strategic modules that generate user value: the repairable click-screen (5" 1080p LCD display (446 ppi) with Gorilla Glass 3 protection), the cameras (front and 8 MP rear) and the rear case (low CO2 emission); and

(B) Commonly used modules, including the midframe with connectors, the antenna module, the lithium-ion battery, the motherboard with its processor chip (Qualcomm Snapdragon 801), the memory (2 GB LPDDR3 RAM) and internal storage (32 GB), the headphone jack, earpiece speaker, and microphone module.

The user-generated repair guide community iFixit gave the FAIRPHONE 2 its first ever perfect score (10 out of 10) for repairability.

For design roadmapping, a quick and enjoyable way to get some basic know-how about a system's architecture is to perform a 'teardown', which means literally disassembling the product. With a deconstructed overview of hardware parts such as of the Fairphone 2 in figure 4.4, you can generate a basic system architecture design. Although a teardown can delve into the tiniest component of a single screw, going into such detail is not necessary for the scouting that aims to provide strategic technologies input for the roadmap. For scouting purposes, the level of aggregation you are looking for is a cluster of components that make up a module adressing a distinctive user value when you renew the module.

One such a strategic module is the screen of the award-winning FAIRPHONE[8]. The screen determines the parameters of its user interface and is also known to be the most commonly damaged component of a mobile phone. Fairphone turned this user insight into the repairability value, leading to a new one-minute screen replacement solution without any tools. Furthermore, the company's future innovation efforts are focused on two other strategic modules that each generate distinctive user value – a camera module, for which the company has plans to launch as an upgrade, and an Near Field Communication (NFC)-embedded rear case that enables quick and easy payments in addition to being made from fair materials with low CO2 emission properties[8].

During your disassembly activity, try not to let the hardware components distract you too much from the software package modules. Increasingly organisations attach strategic importance to software innovations. In the tear down of the FAIRPHONE 2, the main software module, apart from the embedded software programmed in the chips – is its operating system software (Android 5.1 Lollipop).

As stated before, the supply chain markets of components and modules constitute one of the major areas of technology scouting. The others are the areas of technology research and the area of fringe markets and industries that can inspire innovative applications and modules.

Strategic partitioning

When scouting for potential technology options, the strategic partitioning of the system architecture is critical[2]. Strategic partitioning requires that the technical deconstruction of the system allows each module to be coupled to a unique function or feature and also to particular groups of targeted customers[3]. Once that strategic coupling to a module can be arranged, module upgrades of targeted product enhancements within one product line can be realized. When user values can be 'contained' within a single module in the system

TECHNOLOGY SCOUTING & MODULAR ARCHITECTURE

→
Figure 4.5.
Strategic partitioning of the PHILIPS Sonicare dental system. After the example by Ron Sanchez, 2004[3].

cc Joana Portnoy, Anne Brus, Ruben Verbaan & Marco Bonari.
Lecture for the Design Roadmapping Master Course, Faculty Industrial Design Engineering,
Delft University of Technology.

architecture, time-to-market can be accelerated[3].

The technical principle behind the system architecture therefore best follows a one-to-one coupling of function to module[7]. According to this principle, each product function is designed within one module. For instance, the technology roadmap of the PHILIPS Sonicare dental system was based on the strategic partitioning of the Sonicare system and supply architecture as illustrated in figure 4.5[3]. Professor Ron Sanchez showcased this strategic system architecture. He highlighted that Sonicare's roadmapping team followed certain key steps :

→ Deconstructing the toothbrush into functional components, and identifying the seven functional module variations;
→ Determining the strategically needed diversity of the productline range by analysing the consumer new values and newly desired modules first and then comparing this to the competitors provision of key function modules;
→ Partitioning the product architecture into physical components, and identifying the in-house or external supply capabilities required to produce the physical module variations required to provide the desired range of product diversity;
→ Specifying module interfaces in the system architecture to support the product configurability into the strategic range of future product lines[3].

By following these steps, the roadmapping team at Sonicare invented a strategic system architecture composed of six modules - sketched in Figure 4.5 -: (1) bristle unit, (2) housing unit, (3) power unit with integrated drive and battery, (4) printed circuit board unit, (5) charger unit and (6) stand unit. Their technology roadmap plotted the technological evolution of each module. Each technology module was linked to one of seven new product types to be offered in specific design variations covering major retail price points from €15 to €79 over a three-year time pacing horizon, from 2001 to 2003. According to Professor Sanchez, "In essence, the flexibility of a modular architecture to configure a range of strategically desired product and service variations and upgrades happens by system architecture design – as a matter of strategic intent – not by luck[3]."

↘ Team decisions on strategic modules

One baseline decision for designers in mapping the scouted technologies is to decide: which parts, components and modules are of strategic interest to in-house innovation efforts, and which parts are more common and thus easily purchased from a supplier. Effective mapping of the technology system architecture requires designers to work closely with strategic managers. Doing so, they ensure that they can sharpen the definition of their specific goals and ambitions for new business growth, and they also

MODULARITY

VARIABLE RANGE OF COMPONENTS

DESIRED PRODUCT FOR EACH **TARGET GROUP**

SUSTAINABLE (SMART), **FAST** AND **SIMPLE PRODUCT**

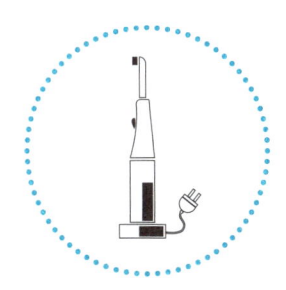

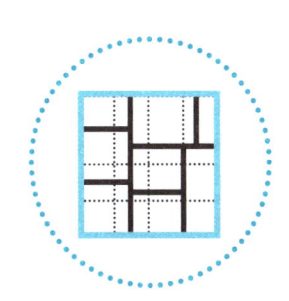

MODULAR PLATFORM

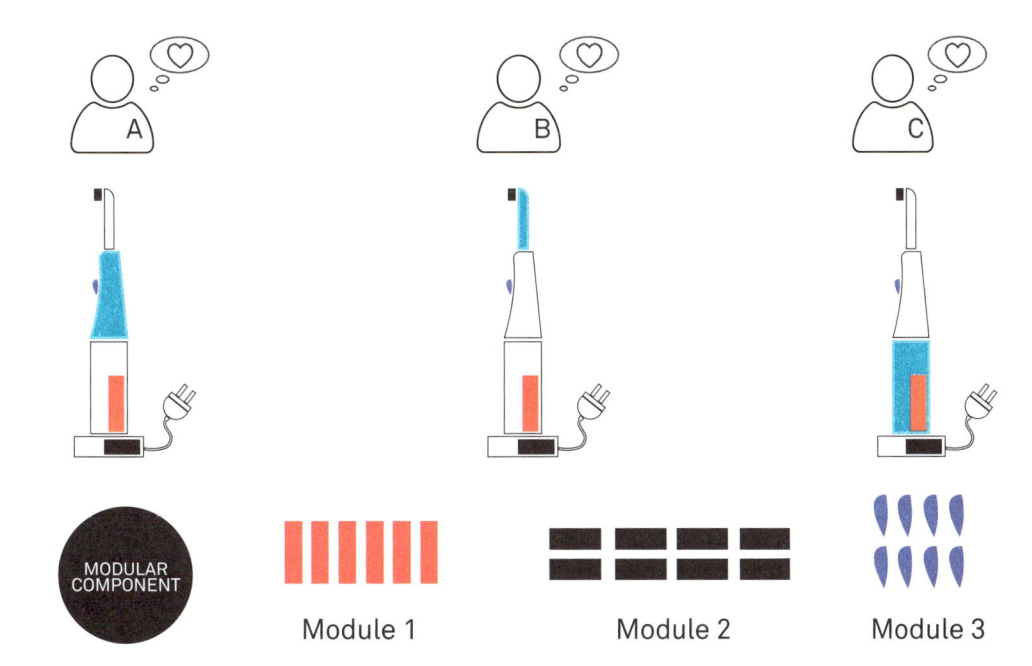

A

B

C

MODULAR COMPONENT

Module 1

Module 2

Module 3

might challenge product (line) managers to adequately formulate their new value strategies for new products in the future marketplace[1].

In roadmapping, the technology scouting is not just concerned merely with the designs for the first generation product or service, but also with the mapping of future generations of products and services. In fact, in technology scouting, extensive collaboration with supply chain managers can have a powerful influence on the future time performance of new products. Close connections in the supply chain enable effective sourcing for complete new module engineering and design[9]. Designers who fully grasp the opportunities and future implications of alternative product and service designs can better support the choice to pursue in-house designs, off-the-shelf designs or strategic partnership designs[9]. Therefore regular interactions with supply chain professionals help designers develop the insight into the most advantageous ways of strategically partitioning the technology roadmap by means of a system architecture and improve aspects of supply chain performance that are strategically important.

Source-based Scouting

Technology scouting relies on formal and informal information sources including the scout's personal network, internet databases and any personal observations made during trade events, supplier visits or shopping trips. Instead of simply executing a randomly generic Google search, setting the scope of the search in advance will help your scouting results benefit[1]. For effective in-depth scouting, it is very useful to think about a list of trusted sources of technology development, drawn from the network relations of the organisation you are roadmapping for. Among the most used sources in technology scouting are (a) supplier sources, (b) patent sources and (c) technology labs.

↘ Suppliers sources

An initial listing of preferred suppliers for your strategically-important modules provides a good starting place for downstream scouting[2]. Besides consulting the websites of the suppliers there are several interesting purchase databases that can lead to valuable search results regarding new components and modules. Most renown is ALIBABA.com. It is also worthwhile discussing state-of-the-art technologies and future options with the managers and technology experts at supplier firms. Tradeshows are another source of valuable information – visiting a tradeshow almost always yields an interesting overview of the markets your suppliers deal with and important observations about upcoming innovations in technology modules and beyond. Make sure to document your observations with photos and notes.

SCOUTING TECHNOLOGIES on formal and informal information **SOURCES.**

↑
Nintendo Switch Teardown

cc iFixit.
Published: March 3, 2017
on https://www.ifixit.com/Teardown/Nintendo+Switch+Teardown/78263. Licensed under the open source Creative Commons.

↘ Patent sources

Patent Listings on the Internet:

→ Espace Patent Database
https://worldwide.espacenet.com/
→ World Intellectual Property Organisation
https://patentscope.wipo.int/search/en/search.jsf
→ US Patent and Trademark Office
http://appft.uspto.gov/netahtml/PTO/search-bool.html

For upstream scouting close to the work of inventors, patent sources are useful[2]. Technology scouts typically use three patent source databases.

Their url addresses are provided above. Besides searches by module and component keywords, we recommend you also to use keywords that include the names of research labs, firms and individual experts.

↘ Technology labs Expert sources

For out of the box scouting of emerging technologies, you will find valuable sources when you turn to experts at technology labs[10]. Well-known labs include the MIT Media Lab, LUCENT Bell Labs, PHILIPS Research and the FRAUNHOFER Institute. Many firms have established research relationships with dedicated labs[10]. For a quick and in-depth evaluation of future opportunities, it is worthwhile to interview several experts working at labs or research institutes. They present an immediate entry point for scouting state-of-the-art technologies. Also, we recommend that you visit a conference in the field of technology pertinent to the strategic modules of your concern.

↘ Scouting reporting with 'One-pagers'

An essential part for pitching the scouting results is the image and description of the technology that explains its application in the business context. The scouts summarize their findings in a one-pager with enough space reserved for an image. Commonly the following types of images are used to explain the application idea of the scouted technology.:

- → Sketches or drawings of patented technology;
- → Pictures of technology demonstrators that showcase the proof of technology;
- → Concept design images, artist's impressions, mock-ups of proof of principles, etc.;
- → Computer-generated images (CAD, rapid prototyping) of a supplier's module;
- → Inspirational images of existing technology applications in fringe areas as an example of modules that might be reengineered, restyled, upgraded, downgraded or isolated.

In addition, typically, the description attributes on the one pagers are:
- → Level of maturity regarding technology performance;
- → User value and business relevance;
- → Estimated costs to either buy or invest in the technology, including technology development defined in terms of man-years.

Figure 4.7 features an example of a one-pagers derived from technology scouting at the printing systems company Océ-Technologies B.V.

OCÉ CASE

System roadmapping
PAUL HILKENS

Founded in 1877, Océ-Technologies B.V. is a global leader in digital imaging, industrial printing and collaborative business services. The company's mission is to accelerate new digital print technologies and transform them into local printing products and services for blue-chip multinationals around the globe and creative studios around the corner. A Canon Group Company since 2009, Océ now operates a vast global network of R&D centres, to connect emerging digital print technologies to future markets. Océ is headquartered in the Netherlands, in the heart of Europe's hi-tech corridor

As Vice President Materials & Device Technology Development at Océ-Technologies B.V., Paul Hilkens is committed to realising innovative system architecture as part of a coordinated technology strategy. This interest began during his Master's degree studies in mechanical engineering at Eindhoven University of Technology, and continues to play a key role in his current position, where he focusses on connecting emerging digital print technologies to future markets

One of his signature projects has been the roadmapping of the platform for the product family launched as the Océ VarioPrint DP line (figure 4.6). According to Hilkens this project is not only evidence of the value of system architecture for organising R&D activities, but equally importantly of the role good roadmapping can play in the presentation and communication of the business proposition throughout the rest of the organisation.

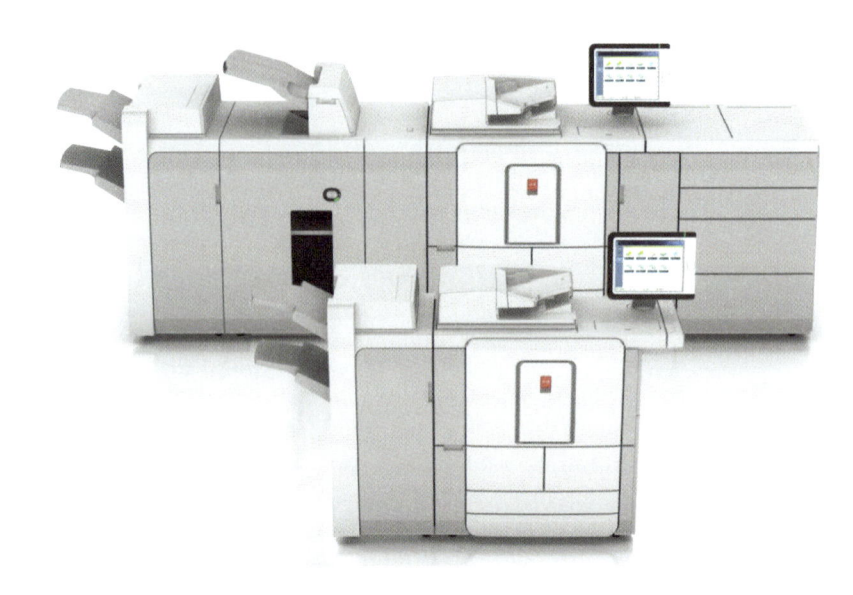

→
Figure 4.6
Océ VarioPrint DP line
Digital printing system for production printing facilities, found in company print rooms and print shops.

© Océ photography.

Hilkens explains how he and his team introduced roadmapping at Océ, and the impact this had: "As a closely-knit team, we worked together for one and a half years. We started by introducing technology scouting alongside system architecture design." Following pilot projects mapping the research and development for two core technology modules, Hilkens and his colleagues created the first platform roadmap focused on a complete product line. He reflects on these early stages of the project: "From the pilots, we knew that roadmapping is a process with an outcome that you don't know when you set off. But at the end of this process there's an outcome that everyone sticks to."

> "When implementing your vision, you put your trust in the fact that having the right people on board will give you the best possible outcome for the future."

↘ Mapping technology evolution

When it comes to mapping technology evolutions, Hilkens is clear about the benefits of combining roadmapping with strategic architecture: "The fact that we are currently seeing new products being launched every year, could lead us to believe that the evolution of technologies and life cycle changes are moving at a faster pace compared with say 10 years ago. But I know from experience that most of these developments have taken much longer than one year; often 2 or 3 years, and sometimes even longer."

> "In order to be able to anticipate change as part of a longer-term strategy, roadmapping offers the perfect way to manage the fit of core technology development with market requirements."

The introduction of the concept of roadmapping at Océ started with a round of stakeholder interviews with senior managers in the organisation. They felt that roadmapping would be strategically important in three ways. Firstly, by involving business strategists, service and business managers earlier on in the development process, roadmaps would ensure new technology innovation efforts were better aligned with the customers' business drivers. Secondly, new system architecture designs create transparency when defining technology development across new printing system development projects. And thirdly, the long-term resourcing of technology evolutions can sustain the in-house development of unique technology capabilities for the future.

When Hilkens first introduced roadmapping to his colleagues, R&D efforts were focused on the physics and chemistry of printing technology – the 'heart' of the printing system – and paid less attention to the functions, features and modules around the core modules. Hilkens explains how this focus shifted: "For the development of this new complete printing system, with unique functions for paper and media handling, we assigned a roadmapping team. Their first job was to carry out full-scale technology scouting research which would provide substantiated input for the roadmap for both the new platform and a new system architecture design."

Time pacing strategy

Previously, when it came to developing new printing systems, Océ had been used to applying an invention-driven approach and project-focused system designs, with minimum relation to module developments in other projects.

Hilkens: "To create the roadmap for the new platform, we felt it was important to map the model year changes, as this rhythm of version upgrades was new to our organisation. We carried out a time-pacing research and investigated innovation cycle times for our own system and those of our competitors. In their roadmap for the new platform, the team set the update rhythm for a new platform generation to seven years, for a significant change to 3 years, and for model updates one year.

Building the system architecture

As the team were working on a completely new platform, the design research for the system architecture included deconstruction of systems used by strategic partners and competitors, as well as those of manufacturing partners responsible for assembling the subsystems. An initial system architecture design was built by a subteam of five system architects (see figure 4.8).

Hilkens outlines the process in more detail: "We held a one-day roadmapping session with the various stakeholders about the setup of the system architecture. We put diagrams on the wall, depicting the function tree of the system and the physical building blocks. The new architecture visualisations generated not only considerable insight about the technologies, but also about the market landscape and the future dilemmas facing our customers in print rooms and print shops."

At the end of just one day, the roadmapping team of engineers, business and service representatives achieved an initial consensus with respect to the building blocks of the system architecture and the creation of a baseline for the modular system, including a view on configurations and upgradeable options. The teams then went on to use this high-level architecture to map out the technology options they had scouted.

↘ Mapping the technology application ideas

The roadmapping team included experts in technology research, competence managers, system architects, a purchasing manager and a manufacturing specialist. All of the team members took on the challenge of carrying out technology scouting. This included desk research as well as carrying out interviews in their own network. The result was a series of long draft lists of potential technology options. These lists provided the input for one of the early roadmapping team meetings.

The team filtered out the overlapping ideas and solutions that had been tested previously, to produce a shortlist of promising technology options. Each team member was asked to prepare a 'one-pager' pitch on one of these ideas and present their findings during the next roadmapping session. The template for these pitches included space for a visual, a short description, a list of the potential customer benefits, estimates regarding the maturity of the technology, the technology application status and availability, cost indications and the source of the information (see figure 4.7).

During a day-long roadmapping session, team members went in rounds to present the shortlisted technology options. The whole team reviewed each one based on its strategic relevance and urgency in relation to future product upgrades. The highest-rated options were mapped on the system architecture block representation. At the end of the workshop, the team had established a raw overview of potential technology applications per module. They had discussed white spots and, where applicable, formulated outlines for new or additional research. -

> "We had never worked like this before. Everyone on the team was enthusiastic about how the discussions had broadened their view beyond their own projects, ideas and expertise. -

The members from R&D particularly valued the feedback from the strategic, service and business disciplines."

→
Figure 4.7
Technology Scouting, 'One-pager' example.

↘
Figure 4.8
System architecture for the platform roadmap of the Océ VarioPrint DP line.

© Guido Stompff, Industrial Design, Océ-Technologies BV.

Technology Description
Change of drive system to be able to deal with a small distance between sheets. It's a package of measures:
- X course with blouse pinch
- High velocity with stopper pinch
- BLDC Xfine motor
- Distance tolerant switch construction (see KM)

Estimated application Status
Product development.

Cost indication to next milestone (FTE/Investment)

	Time	MY
2M1	0	0
2M2	12	4
2M5	12	2

Potential Benefit for Customer
Higher PPM to counter speed creep and increase productivity.

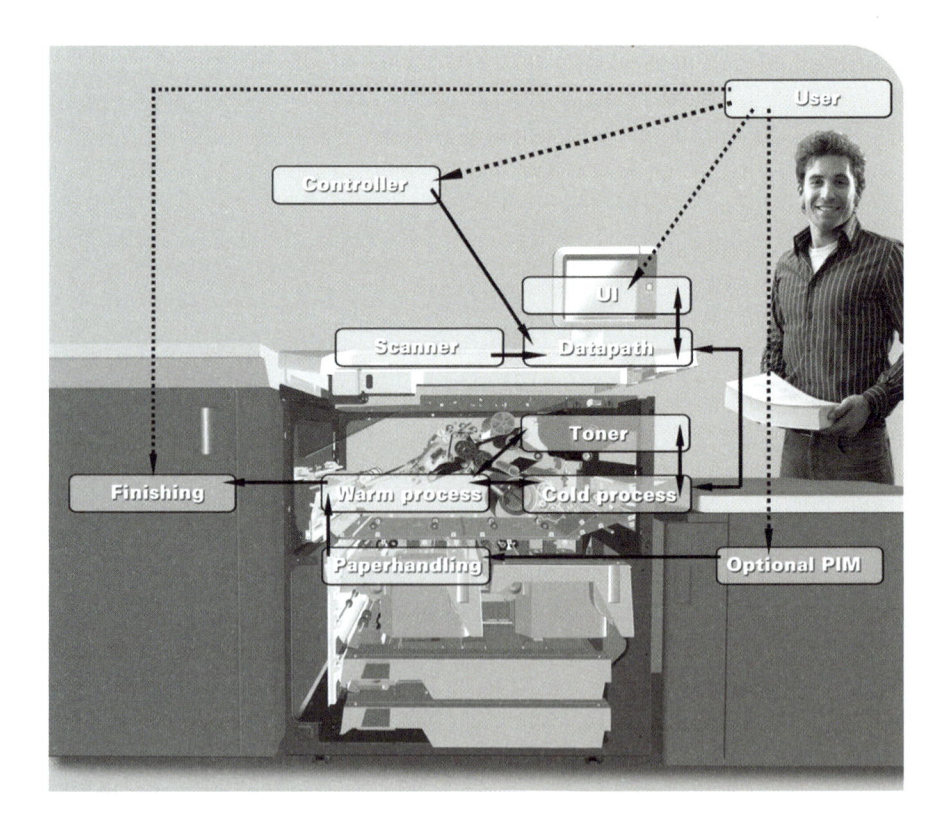

↘ ## The platform roadmap

The specific strategic challenge of the platform roadmap (see figure 4.9) was to overcome a situation that a product line has as many different configurations as it has customers. The scouting research carried out for the system architecture was used as a basis for addressing this challenge. A subteam of five multidisciplinary architects built an initial system architecture design. The vision of the roadmap was to provide a range of customers with an upgradable series of configurable printing systems. For customers with a lower budget, one system unit was designed. And for the high-end segment, the team came up with a modular product configuration of 10 - 12 configurations.

The upgradable configuration propositions became the leading theme for the innovations mapped for the first 1-3 years. For 5-7 years and beyond, the theme was 'Inkjet is the future for our new colour technology'. Discussions in the roadmapping team focused on the years in between (3-5 years). The strategic guideline for these decisions was to map only those new modules for the black and white system that could also be used, or reused, to build a potential colour platform. "The complex nature of printing system technologies is such that you have to adopt a broadminded approach, and cannot simply base your innovation strategy on the opinions of a single local hero. In order to create a complete image of the future, you need the contributions of the best experts in the organisation. The ones who have the combined knowledge and experience

– making sense of this complexity by mapping the evolution of the system in a coherent and complete way."

Hilkens can now look back on a successful platform roadmap. Since the resulting system, the Océ VarioPrint DP line, was launched in 2010, thousands of units have already been sold and continue to be sold to professional printing customers. In a highly competitive market segment, the team managed to achieve a unique value positioning, thanks to the technology asset of upgradable system configurations, smart service anticipation and reliable control of productivity. Hilkens can be justifiably proud of the strong position the first Océ roadmapped platform now occupies within the global Canon portfolio.

PAUL HILKENS MSc, is Vice President Materials & Device Technology Development at Océ-Technologies B.V. He began his career with the company as a young mechanical engineering graduate in 1996. Over the years, he has held positions as a multidisciplinary researcher, function designer and system architect, before becoming Head of System Development in 2007. In this position, Hilkens was responsible for introducing roadmapping for system architecture development. In 2009, he was promoted to Vice President R&D. In that role, he was subsequently responsible for the technology areas of electronics and embedded software, hardware and industrial design, and is currently responsible for materials and printing technologies development. Within the Canon network of R&D centres, Hilkens operates globally, to advise colleagues and realise leading-edge technologies for professional printing customers around the world.

LAB ↗

Tear down the system you are roadmapping to establish a system architecture

MATERIALS NEEDED:
- → screwdriver
- → camera
- → big sheet of paper
- → note book, pen
- → laptop

1 To begin, get a hold of the current version of the device or system you are making a roadmap for and prepare to take it apart at your workplace.

2 Take a screwdriver, or whatever tool is appropriate and dismantle the device part by part. Exercise caution during disassembly. Inspect all the parts.

3 Keep track of each step of the disassembly and take pictures and notes as you work. Most importantly, take a photo featuring the complete teardown, with all the components neatly organised.

4 Use the teardown shot to add images or drawings of software packages that have been programmed in-house or that have been bought off the shelf.

5 Use the teardown overview for establishing the system architecture. Decide which parts are strategically interesting for users and consumers and which ones are common and hence easily purchased from a supplier.

6 Cluster the user value parts of both hardware and software into modules that can be developed as incremental upgrades.

7 Cluster the common parts into modules for suppliers. Tag each cluster with a module name, and generate keywords for the module to inform and direct technology scouting efforts.

Prof.dr.ir. PASCAL LE MASSON on the subject of Strategic Design and Industry roadmapping

LS   Your scientific research made a contribution to positioning strategic design[11]. What do you consider important with respect to strategy as 'innovative design'?

PM   An innovative design perspective on strategy rests on the generation of alternatives and implies entering a 'post-decisional' paradigm. Before any decisions or actions are undertaken, strategic designers aim to design a 'decision space' that helps to go beyond the strategic dilemmas that organisations face. They overcome the issue of strategic fixation and provide a creative capacity to think around it. It relies on the theoretical foundation of contemporary, advanced design theory - like Hatchuel and Weil's C-K theory[11] – as well as solid cognitive studies at the experimental and ecosystemic level. The 'decision space' also has organisational consequences: since the design of alternatives becomes clearly a strategic matter, linked to innovation department and innovation organisations.

LS   In your experience of doing research in the semi-conductor industry, how can designers deal with the challenges associated with fast-paced technologies?

PM Together with my colleagues, I worked in the semiconductor industry for quite some time. It is a fascinating field of research, and a leading industry in terms of roadmapping. Their technology sits at the forefront of innovation with a constant renewal of techniques, the markets are always evolving, and so designers are constantly confronted with a 'double unknown': unknown markets and unknown technologies. The semiconductor industry teaches how to deal with that double unknown by applying a two pronged logic[12]- that at first glance appear counter-intuitive (but is highly rational!):

COLLEGIAL MANAGEMENT OF THE UNKNOWN: At the ITRS - International Technology Roadmap for Semiconductors - organisation, industry leaders explore the unknown future of semiconductors collectively. In a way that is strictly organised, with clear rules and administrative forms to follow. These rules are actually 'unlocking' in the sense that they constantly push the actors to explore the terrain that lies beyond collective fixation[12].

DESIGN OF GENERIC TECHNOLOGIES: innovators in semiconductor companies do not only think on the level of products and users, they also seek out at the level of unknown technologies that could be generic enough to be relevant for several markets – generic not as a random walk of trials and errors and gambles on markets. It should be the result of an organised process that supports generic design outcomes[12].

LS What is your opinion on creating modular system architectures for innovation strategies and roadmaps?

PM You are completely right about the power of modularity. Baldwin and Clark[13] remind strategists that there are several ways to address a set of requirements, among these modular solutions are best. A modular system architecture enables a company to address mass-variety by combining modules for mass customization. Providing an adequate response to uncertain market demands and remain robust in fast-paced environments. Hence the technique to structure the system is critical in leading to interesting strategic positionings, such as platform leader or event complementor. The burning questions today are: How to get to these positions? How to design a platform? How to organise a collective whose aim is to design a platform?

LS Also critical to this is technology scouting and generating promising applications. This could for instance provide input for a platform roadmap. Which scouting sources do you recommend?

PM Concerning strategy and creating 'decision space', the goal is to overcome fixation and this question raises the issue: Which sources will help roadmapping teams overcome it? There are so many knowledge sources available in today's world of data and open innovation – Designers should learn to make use of data science and data intelligence to fuel surveys, networking, open innovation, contests and a wider involvement in scientific and design communities. But that still isn't enough! As we know from design theory, new knowledge also comes from new concepts: Having a creative imagination, and a capacity to formulate 'crazy concepts' is important, because these are the kinds of ideas that lead to 'crazy' questions about who can provide the appropriate knowledge, data or solutions. Designers should get to know about the 'crazy' ideas in their fields by visiting design schools, design studios and any other place where imagination is at work! Designers can also use 'smart browser' enhancements that suggest the key terms they need – but may not be familiar with – to investigate a certain topic[14]. Even a smart web browser can support absorptive capacity! Designers would do well to keep these kinds of tools in mind .

LS My last question is: what is your advice to design roadmappers who are mapping new technology applications to product service systems?

PM Be rigorously creative. Design roadmapping is the capacity to systematically formulate every imaginable future. Thanks to contemporary design theory we know that this work can be carried out rigorously, and thanks to advances in cognition we know that without rigour, we tend to become fixed. This is why we need processes, techniques and guidelines for organisations, like the ones described in this book!

PASCAL LE MASSON is Professor at MINES ParisTech –Paris Sciences et Lettres Research University. where he chairs the design theory and methods for innovation group and is also the deputy director of the Centre for Management Science. Pascal is Head of the engineering design curriculum of MINES ParisTech. In the international community of scientific research on design innovations his positions include, chairman of the 'Innovation' special interest group of the European Academy of Management and chairman of the 'Design theory' special interest group of the Design Society.

PLATFORM ROADMAP

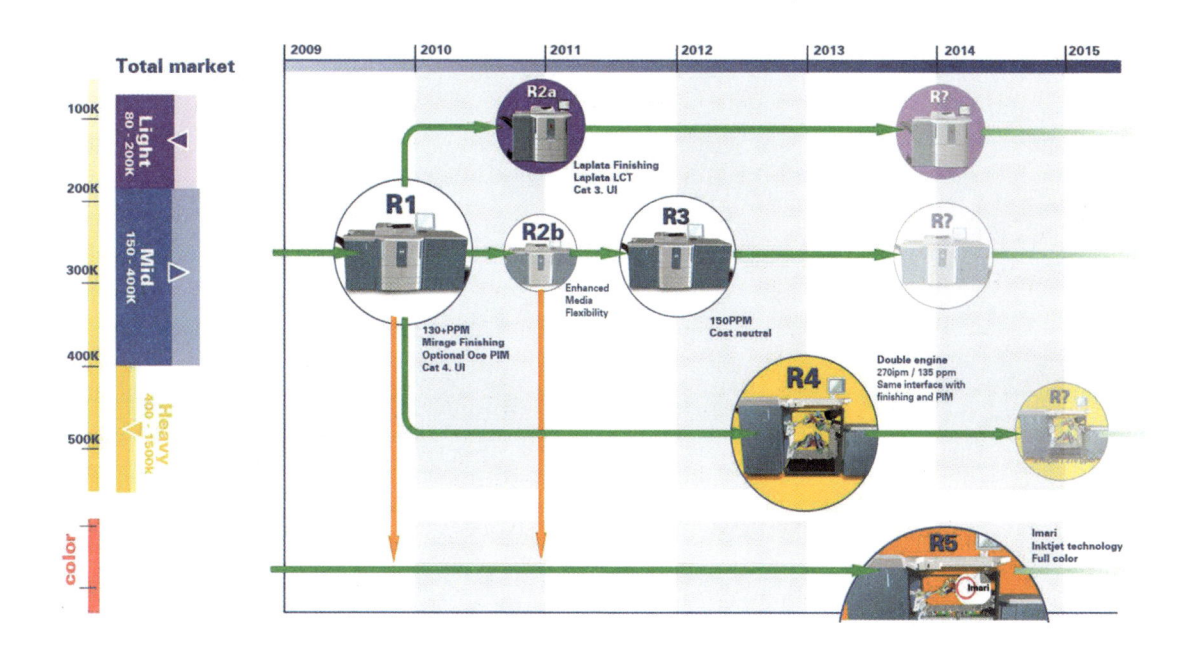

Figure 4.9
Platform Roadmap
Océ VarioPrint DPLine
- *Paul Hilkens.*

© Guido Stompff, Industrial Design, Océ-Technologies BV.

↗
Kinetic Plastic Heureka,
Jean Tinguely, 1967 Zürichhorn.
Detail shot.

cc Ogre Bot, photography.

IN SUM

Technology Scouting includes a higher level of detail than simply tracking technological trends. In this chapter, the focus was concentrated on the usefulness of system architecture. Modular architecture provides the structure needed to map technology evolutions and also serves as the radar for the technology scouting activity.

We gained knowledge on:

- → The mapping of technology evolutions such as in lighting and display technology;
- → The difference between out of the box scouting and in-depth scouting;
- → The usefulness of a tear down to generate a basic system architecture; and
- → The different types of sources to include in a technology scouting research.

As guidelines, we have provided you several examples, a lab, a case story by and a science interview with Pascal Le Masson on the practice of mapping technology evolution, generating a modular architecture and source-based scouting.

---

1. Phaal, R., Simonse, L.W.L. & Den Ouden, E.P.H. (2008). Next generation roadmapping for innovation planning. International Journal of Technology Intelligence and Planning, 4(2), 135-152.
2. Sauer, A., Thielmann, A. & Isenmann, R. (2016). Modularity in Roadmapping–Integrated foresight of technologies, products, applications, markets and society: The case of "Lithium Ion Battery LIB 2015". Technological Forecasting and Social Change. available online from 27 August 2016.
3. Sanchez, R. (2004). Creating modular platforms for strategic flexibility. Design Management Review, 15(1), 58-67.
4. Albright, R.E. & Kappel, T.A. (2003). Roadmapping in the corporation. Product–technology roadmaps define and communicate product and technology strategy along with a longer, smarter view of the future. Research Technology Management, 46(2), 31-39.
5. Fusion Imaging 3D printing of Drone parts. https://www.shapeways.com/shops/fusionimaging accessed August 2017.
6. Nest's self-learning thermostat https://nest.com/thermostat/meet-nest-thermostat accessed August 2017.
7. Ulrich, K. (1995). The role of product architecture in the manufacturing firm. Research policy, 24(3), 419-440.
8. Modularity strategy goals of Fairphone https://www.fairphone.com/en/our-goals/ accessed October 2016.
9. Caetano, M., Amaral, D.C., (2011). Roadmapping for technology push and partnership: a contribution for open innovation environments. Technovation, 31(7), 320–335.
10. Rohrbeck, R. (2010). Harnessing a network of experts for competitive advantage: Technology scouting in the ICT industry. R&D Management, 40(2), 169-180.
11. Le Masson, P., Weil, B. & Hatchuel, A. (2010). Strategic management of innovation and design. Cambridge: University Press.
12. Le Masson, P., Weil, B., Hatchuel, A. & Cogez, P. (2012). Why aren't they locked in waiting games? Unlocking rules and the ecology of concepts in the semiconductor industry. Technology Analysis & Strategic Management, 24(6), 617-630.
13. Baldwin, C.Y. & Clark, K.B. (2000). Design Rules, volume 1: The power of modularity. Cambridge MA, US: The MIT Press.
14. Kokshagina, O., Le Masson, P. & Bories, F. (2017). Fast-Connecting search practices: On the role of open innovation intermediary to accelerate the absorptive capacity. Technological Forecasting and Social Change, 120(7), 232-239.

TIME PACING STRATEGY

TIME PACING STRATEGY

HOW TO DECIDE ON THE TIMELINE INTERVALS

In history class, timelines are introduced to enable storytelling about development and evolution. In roadmapping, the timeline design has a similar purpose, but instead of concentrating on the historic past, the timeline addresses the future. The timeline on the roadmap allows designers to create stories about development and evolution of future products and services.

Time in the classic period has been a favourite subject among philosophers. Aristotle talked about chronos as the sequence of time, the passing of time. He defined chronos as the 'number of motion in respect of before and after' in Physics 219B (IV xi)[1]. His notion of time can be measured, is dynamic, and has a chronology, a timeline. Plato, on the other hand, drew attention to time as kairos: a moment of opportunity in which choices can be made, the opportune time. Kairos, according to Plato, is the right time when something of importance can happen based on decisions concerning future action[1]. Both notions of time come together in roadmapping. The timeline in a roadmap focuses on the future chronology of chronos and the pacing of kairos opportunities of innovation. Rather than linear stories, roadmapping favours parallel stories about future scenarios that connect to the vision.

In the roadmapping undertaking, you need to decide upon the length of the timeline, how far the vision will be projected to a future point in time, and over what time intervals new design innovations are mapped. The scope of the roadmap, its industry context, and its strategic innovation cycles are important parameters in this decision.

In this chapter, the design of the timeintervals for the foresight of innovations is guided by three decisions on the strategic horizons and the moments of transition, the designed rhythm of innovation—the design clock—and the time performance.

Horizons of Strategic Life Cycles

↘ Three Horizons

Of particular help in parallel storytelling is the futures technique of the Three Horizons model[2]. This technique comprehends three parallel scenarios based upon three different life cycles of strategic business innovation. These life cycles overlap, as modelled in figure 5.1, to create continuous innovation on the long term[3]. Each life cycle conceptualizes new business development. The first starts in a current business with existing market and existing technologies and concentrates on innovations of design value enhancements. The third, projects new value propositions in a new market with new technologies. The cycle in-between is the stepping stone towards either a new market by new user value segmentation or by totally new technology applications that are user-tested in an existing market. Ideally, these major strategic cycles are managed with a continuity of smooth cycle transitions, to maintain continuity in turnover[3].

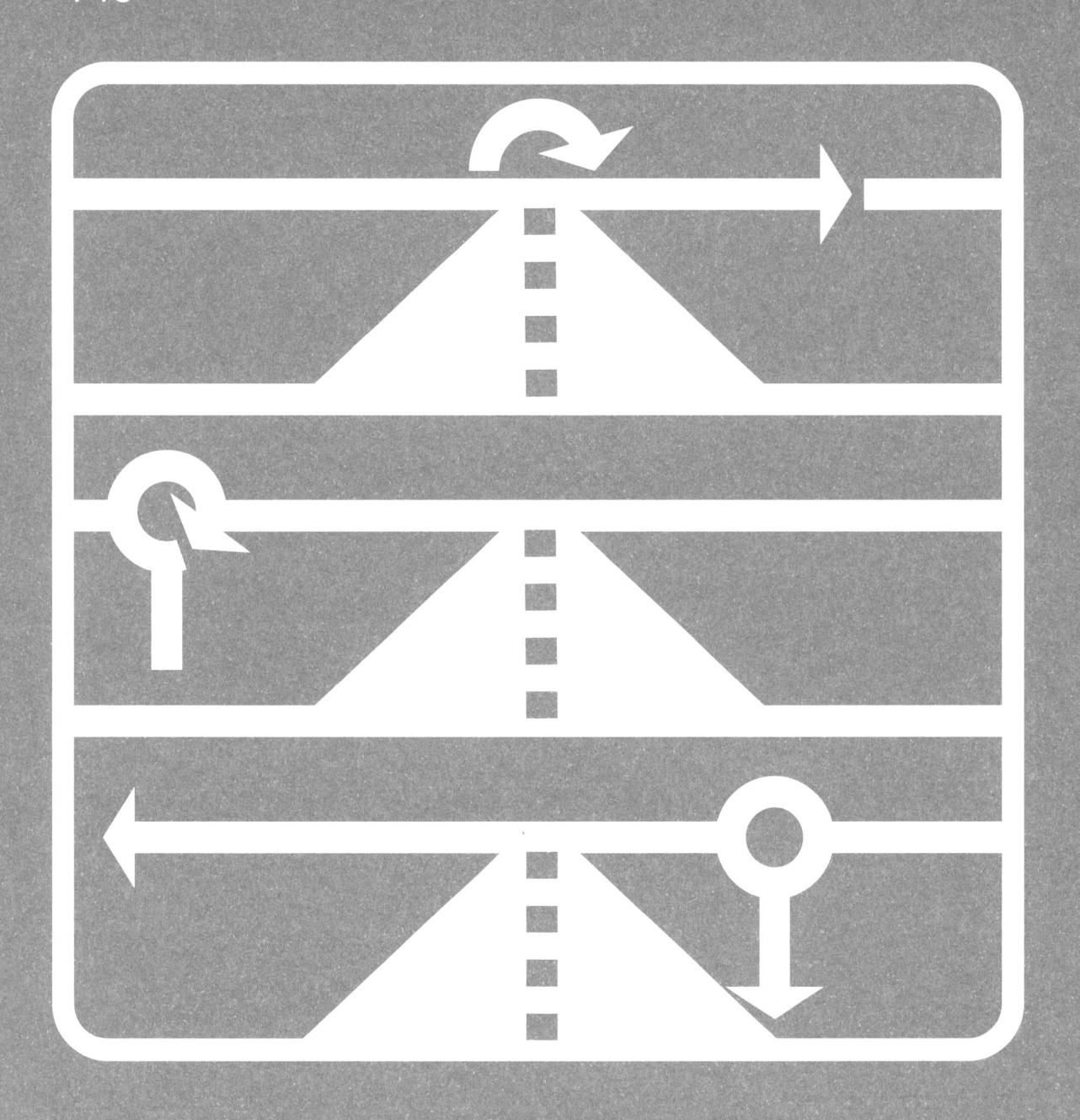

HORIZONS mark transitions of the strategic life cycles of business innovations.

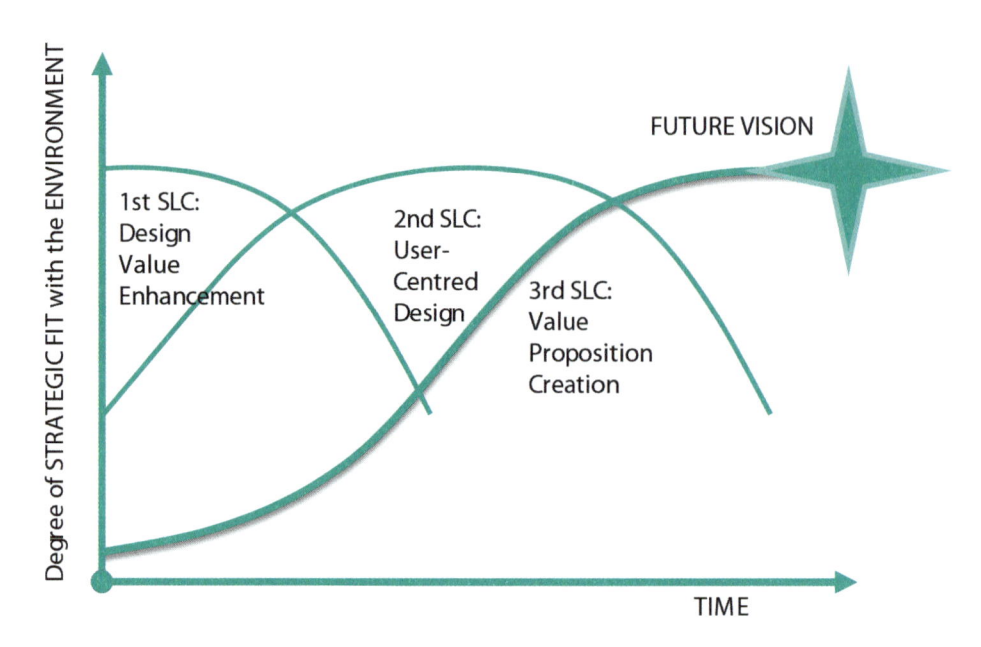

Measured Life cycle curves

The schematic curves of strategic life cycles have also been measured in sales data. Figure 5.2 shows a reality check of the life cycle curves, measured by the Motorola roadmapping team[4]. The measured curves represent the actual product life cycles of car audio systems and resemble the schemtic ones quite closely, although they are a bit rougher and bumpier. Overall, the stages of growth, maturing, and decline are clearly recognizable.

Motorola's roadmapping team measured the units of sales of different products in this car audio product line over a period of several years[4]. The first two data curves (A and B) in Figure 5.2 indicate that a sales dip occurred between product A and product B. The timing of the market launch of product B was a bit too delayed to enable a smooth transition in sales. The transition between products B and C was much better timed, while the transition between products C and D was excellent. For Motorola's roadmapping team, these historical data analytics were important in deciding on the time pacing strategy of the roadmap[5]. The team determined the starting point for the development projects of the next car audio products (H and I) by estimating the number of months before market introduction. Roadmapping smooth cycle transitions, is a cornerstone of the time pacing strategy[5]. When you start with roadmapping, doing a similar analytical exercise can be full of insights. You can analyse the past innovation cycles and project them onto the future timeline.

Figure 5.1
Strategic Life Cycles model of Three horizons.

cc Simonse & Hultink, 2017[3].

Figure 5.2
Motorola Life cycle curves show past product performance history, of units of sales, in order to project new product life's

cc Willyard & McClees, 1987[4].
Motorola Inc

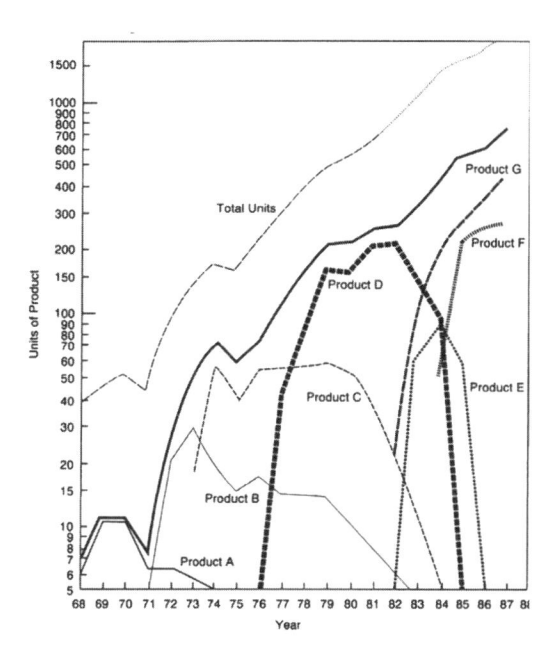

Three strategic life cycle scenarios

The 'three horizons' model focuses on three strategic life cycles projected to a future timeline[3]. The first concentrates on design innovation in a business that will mature over time (the current businesses). The second cycle envisions compelling new user value segmentations and transforms the current business into the newly envisioned business that follows into the third horizon of the disruptive new value proposition creation of a highly promising upcoming business. The scenarios are represented by strategic life cycles, as depicted schematically in figure 5.1.

HORIZON 1 → Design Value Enhancements
The first horizon envisions a strategic scenario of a continuous flow of enhancing design value to current product- or services lines. These value enhancements are for instance model changes in shape, colour, or extra features. Essential in this strategic scenario, is the reuse of existing modules and functions. Commonly, around eighty percent or more of the prior product remains unchanged. It's time pacing is focused on bringing new versions to the market by a certain update rhythm[5]. The design innovation effort continues the product line through maturing stages of evolution, until the moment arrives that signals a decrease in sales and profit. Eventually this scenario includes the strategic situation of decline in which the business will gradually lose strategic fit with its environment. The endpoint of the strategic life cycle of this business is

within view[3]. Innovation practice has shown that it often takes longer to accept the reality that a business is declining than would be rational and best for the company[6]. Apparently, it is quite difficult to change the mindset of people who have become accustomed to successful products. It is not uncommon that a roadmapping process starts in a situation of a business that loses strategic fit.

HORIZON 3 →   Value Proposition Creation
The third horizon captures a strategic scenario with a state of growth on the long run. It is a disruptive innovation scenario with a new value proposition that inhibits the potential to displace the system of the first strategic life cycle. The future vision we discussed extensively in the third chapter, is the end point of this innovation scenario. The desired values of the imagined value proposition of the future are, at best, marginal in the present. Rather than a progression, this third horizon's scenario inhibit a disruptive change of new value(s) over time that potentially offers a more effective response to the external environment[3]. The time pacing is expected to be more long term, taking several years for the new business development of new value propositions that also might involve the creation of new business models[5]. The start of this strategic innovation scenario is in the present by seeking a strategic fit with emerging signals.

HORIZON 2 →   User-centred Value Creation
This strategic scenario of user-centred value creation in the second horizon falls between the disruptive and the enhancement scenarios. It concerns growth and transformation and thrives on design research dedicated to insights about the desires and dilemmas of users[3]. New insights on emotional and functional values allow designers to create new markets with new products and services that differentiate from the existing product/market combination. Such a newly discovered value insight delivers new meaning for a new market segment[5]. Furthermore, the design of a next generation product, platform or service that incorporates the application of new technologies can be part of this scenario[3]. The user-centred value creation includes then the testing of the new technology application by users in the existing market. A major challenge in this strategic scenario is overcoming the dilemmas for reaching the third strategic life cycle. In this intermediate time space, the current and envisioned product lines collide. This scenario is therefore typically unstable and characterized by clashes of multiple values and directions of creative solutions[3]. It takes an entrepreneurial mind-set to identify propositions that enter the growth stage of critical market acceptance. Making sense of the lessons learned about user acceptance of new technologies is crucial for this scenario[3]. Therefore, in this scenario, several alternative paths of value

→
Figure 5.3
Kickstarter case study Kamibot

cc Joana Portnoy, Anne Brus, Ruben Verbaan & Marco Bonari, 2016. Lecture for the Design Roadmapping .Master Course Strategic Product Design Faculty Industrial Design Engineering,
Delft University of Technology.

TIME PACING STRATEGY

KAMIBOT CASE STUDY — Three Horizons

KAMIBOT CODING 2D PAPER	KAMIBOT CODING 3D SHAPE CLOUD	KAMIBOT KIT CODING 3D DRAWING KAMI-MIND
+1 year	+3 year	+10 year

2D Shapes:
- Develop shapes
- Design store

Coding:
- Children coding program Software, interface, tutorials

3D Shapes:
- Develop shapes
- Design store

Cloud:
- Develop platform for cloud
- Acquire database for cloud
- Mantain and manage it

3D Drawing:
- Children 3D-drawing program software, interface, tutorials

Kami-Mind:
- Personality Software
- Learning Software
- Reacting Software

Design Kit:
- Make it
- Make tutorial

VARIABLE: COVER, PAPER SKIN (DESIGNED), PAPER SKIN (IN BLANK), SOFTWARE

COMMON: IR SENSORS, ULTRASOUND SENSORS, MOTORS, RGB LED

PRODUCT:
- KAMIBOT
- KAMIBOT 3D
- KAMIBOT KIT
- KAMIBOT KIT 2.0

SERVICE:
- LEARN TO CODE
- 2D MODEL STORE
- SCHOOL SUPPORT
- 3D MODEL STORE
- KAMI-CLOUD
- KAMI 3D DRAWING - Design your own robot
- KAMI-MIND

TECHNOLOGY:
- CLOUD-COMPUTING
- 3D SHAPES
- NEURAL NETWORK
- SOFTWARE for 3D DRAWING
- EXPERIENCE VR
- 3D PRINTER for KIDS - Plastics

| +1 year | +3 year | +10 year | Vision |

↑↗
Real Time Schiphol, 2017.

© Maarten Baas
Courtesy of the artist
Rob Hodselmans photography

"Real time is a term that is used in the film industry. It means that the duration of a scene portrays exactly the same time as it took to film it. I play with that concept in my Real Time clocks by showing videos where the hands of time are literally moved in real time" - *Maarten Baas*.

Real Time clocks show a video performance made by BAAS which takes exactly twelve hours to film and twelve hours to watch it in its entirety, thus creating a hyper-realistic representation of time.

TIME PACING STRATEGY

creation are pursued at the same time until it becomes clear that one of them has been accepted by a critical mass of users and the unsuccessful alternatives can be discontinued.

On a roadmap, all three strategic life cycle scenarios can be mapped to meet simultaneously the short, mid-term, and long-term business development of strategic fit with its future environment. They offer three parallel options for design innovation efforts projected on the future timeline of the roadmap[3].

Venture case example

To illustrate the three horizons thinking, a team of design students developed a case example based on the venture Kamibot (see figure 5.3). Kamibot is a venture that makes programmable papercraft robots for kids and has raised capital on the crowd funding site Kickstarter. The Kamibot robot teaches kids how to code in a fun way. They can customize their toy with their own code and colourful paper skins.

→ For the Kamibot venture, the first horizon scenario of incremental innovation involves updating the robot. To prepare it for future generations, the team considered that it is crucial to enhance it with coding software for children.

→ For the third horizon scenario, the team created a vision for Kamibot: 'Empower children to design and program their own toys'. As driving values in this vision, they defined (a) stimulate creativity, (b) learn 3D design, and (3) make the toy more personal. To enable this, the team selected promising technologies: 3D programming for kids and the launch of a new service for drawing your own robot in 3D.

→ For the second horizon scenario, the innovation concerns a next generation of Kamibots that makes use of 3D-printing technology for kids. This scenario includes the path of a new service: of a 3D model store. Another alternative path of innovation that will also be tested in this second horizon is the introduction of robot learning.

The team translated the scenarios of the three horizons into choices that captures the main product, service, and technology elements by plotting them on a 10-years-timeline on the roadmap as illustrated in figure 5.3.

Design Clocks

Transition points on the timeline

On the roadmap, the envisioned launch of a new product or service is often demarcated with a specific point in time that relates to an

TIME PACING STRATEGY

DESIGN CLOCK is the designed RHYTHM of innovation by modes of design innovation.

event or exhibition[5]. For instance, for roadmappers in the fashion industry, Paris's Première Vision event in February is important. For the bike industry, the Eurobike show is the gathering place of industry players, press, public, suppliers, and competitors. And for consumer electronics, these crowds gather at the global consumer technology tradeshow CES, which takes place every January in Las Vegas. These events mark the transition point in times from old products to new ones.

The new products and services are announced and shown to the press and public in these market arenas. When deciding at which event a product should be launched, roadmapping teams take seasonal transitions into account. For instance, PHILIPS' roadmapping teams choose launching in January at the CES to prepare for the availability of their products for Valentine's Day, spring, and Easter. And launching in the autumn, in September at the IFA, the trade show for consumer electronics and home appliances in Berlin, so that the new consumer electronic product will be available in retail channels before Black Friday and Christmas. The exhibition month is designated as the future point in time on the roadmap. These events offer a 'hard' deadline for the delivery of future innovation projects.

↘ Time pacing strategy per design clock

The time pacing strategy[7] decides the 'design clock' for the different types of design innovations per horizons. Design clocks determine the rhythm of innovation efforts by regular deadlines—time stones.

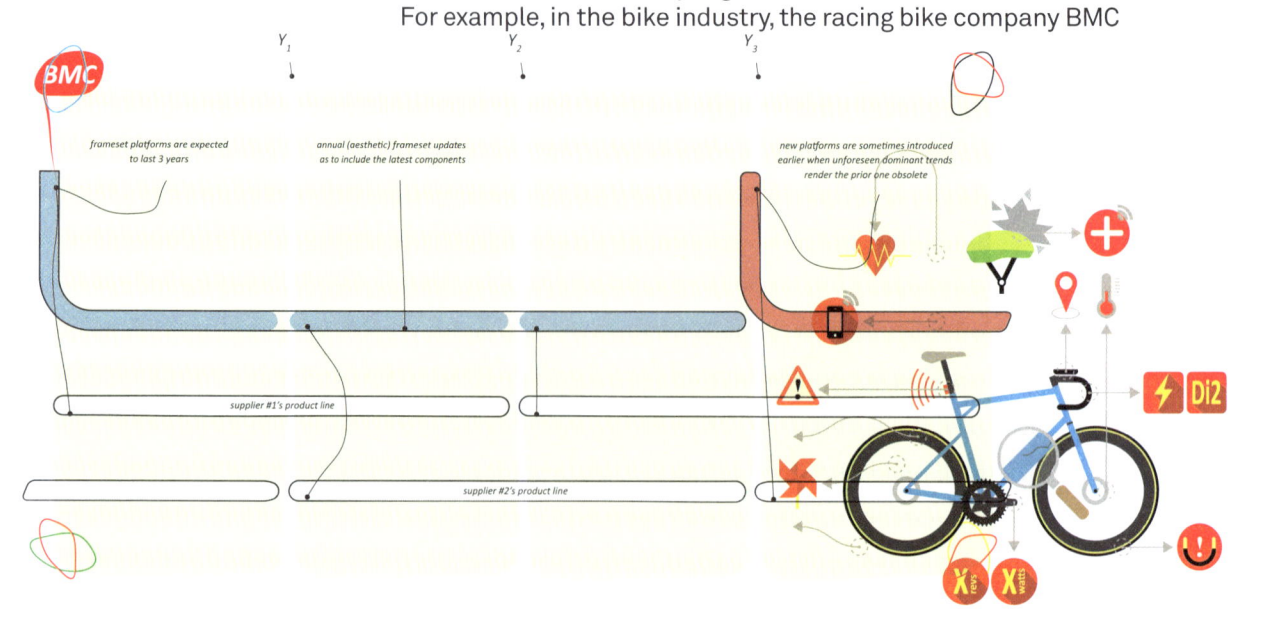

For example, in the bike industry, the racing bike company BMC

Figure 5.4
D-Bike project BMC

cc Wouter Aerts, 2016.
Master Strategic Product Design, Graduation project.
Faculty Industrial Design Engineering,
Delft University of technology

Figure 5.5
Value creation by three modes of design innovation.

Simonse & Hultink, 2017[3].

releases a new bike each year. Many competing bike companies use the same release rhythm of annual 'model year changes'. In addition BMC also programs new platform innovations in which the same frame (the platform) accommodates a sequence of new module innovations. The time pacing used for launching every next platform innovation is three years. In the first year, BMC introduces the next generation bike. During the subsequent two years, they enhance the frame with module innovations. A third time pacing strategy plots the vision of a radically new type of bike, such as the digital racing bike, the D-bike, that integrates Internet of Things technology to provide completely new user services (see figure 5.4). The three time pacing strategies are associated with three design clocks. The design clock for the 'model year change' is one year, for new platform generation it is three years, and for radical innovation it is about ten to twelve years.

Gazelle, a Dutch bike manufacturer, has similar design clocks as BMC for the model year changes of its racing bikes. However, for lifestyle bikes that people use for getting around in cities, the company decided to take a slower pacing of the innovation efforts, with a two-years design clock for model changes. They based this decision on differences in the user target groups of racing bikes and lifestyle bikes. The techy user group interested in racing bikes has much more interest in the latest performance-enhancing innovations. These bike enthusiasts read forums and magazines to keep up with the latest developments and they are therefore likely to replace their bike more often than the users of lifestyle bikes. The lifestyle bikers appear to be more interested in safety issues, comfort, and styling for which changes over a two-year design clock, seems to better match.

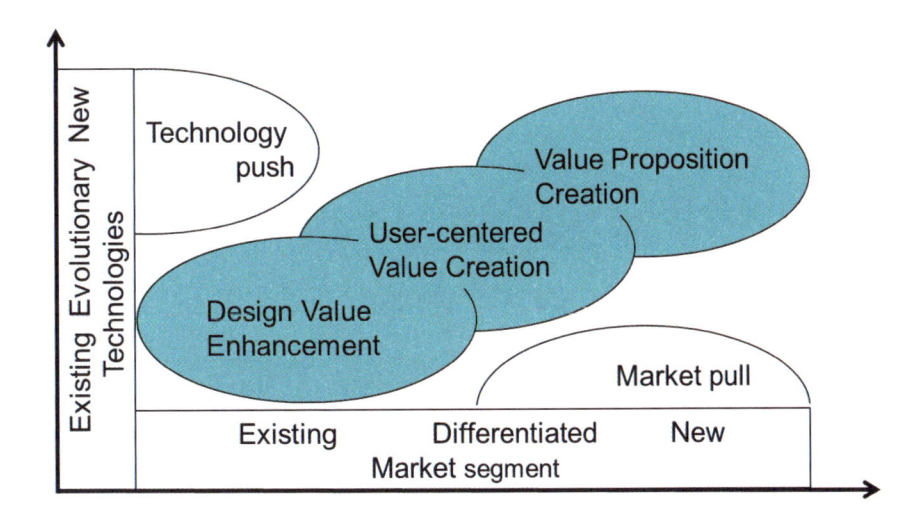

In figure 5.5, three modes of design clocks are modelled by two dimensions of technology and market innovation. The model presents three modes of design value that bridges the market pull and technology push, bringing these two forces of innovation together[3]. The design clocks guide the time pacing of these value creations by design:

→ DESIGN CLOCK OF VALUE ENHANCEMENT
The value enhancement of model year changes often includes a new shape, colour, or extra features, each of which enhances the design value of the product or service. This type of design innovation thrives on incremental changes on a product line or service family positioned in an existing market with existing technologies. Its time pacing is fast, focused on bringing a new version to the market. The update rhythm accommodates this fast speed of innovating by design value.

→ DESIGN CLOCK OF USER-CENTRED VALUE CREATION
User-centred value creation thrives on design research dedicated to insights about the desires and dilemmas of users. New insights on emotional and functional values allow designers to create new market segments for new products and services that differentiate from the existing product/market combination. It's design clock relates to a new market segment established by a newly discovered value insight that delivers the new meaning. Its time pacing is moderate as this type of value creation concerns a next generation solution design that also might incorporate the application of evolutionary technologies. The update rhythm of new user-value solutions often takes a number of years.

→ DESIGN CLOCK OF VALUE PROPOSITION CREATION
The value in value proposition creation encompasses a broader impact, including the new organisation design of the value chain. Creating new value propositions often requires a new business model that integrates new technologies with new markets. The time pacing is therefore much longer, taking a rhythm of several years for new business development.

The three 'design clocks' for different modes of design innovation offer roadmappers options for the time pacing of different design innovation efforts in the roadmap[3]. For the rhythm of new products or services, time stones can be mapped on the timeline of the roadmap[5].

Time Performance

ERIK-JAN HULTINK AND LIANNE SIMONSE

↘ Competitive timing

Deciding on how many years ahead a next generation product or service should be paced also depends on the competitors in the

TIME PACING STRATEGY

Measuring the launch PERFORMANCE of product visions with a particular future time point.

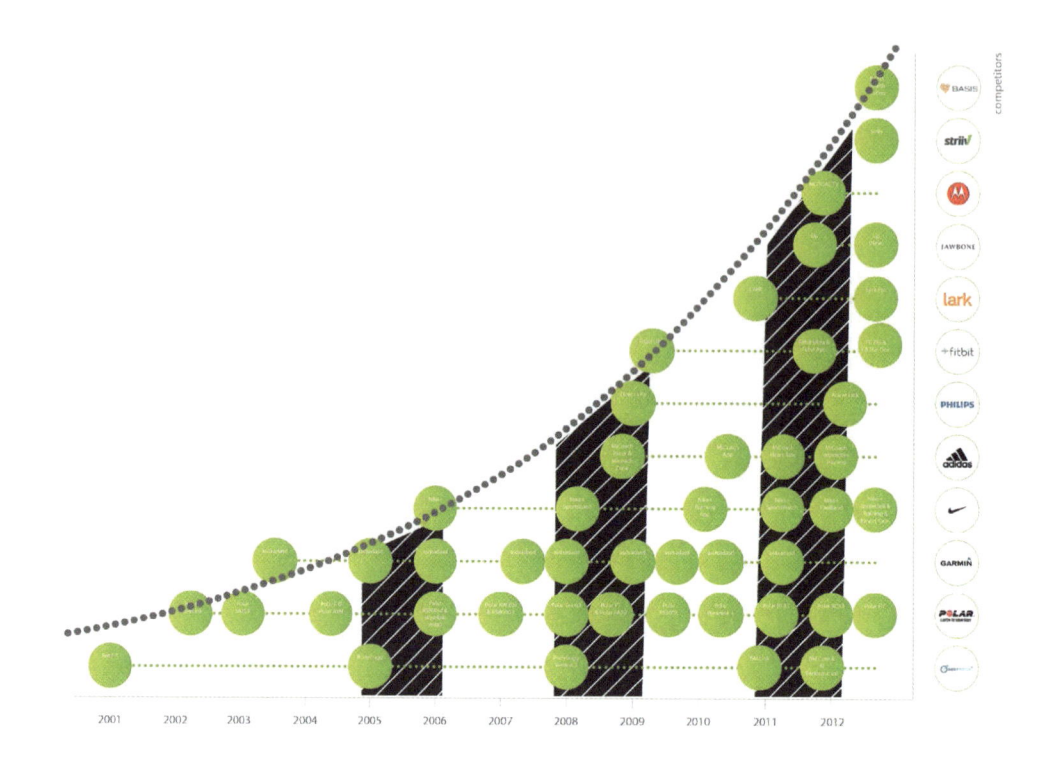

marketplace[5]. Figure 5.6 shows the emerging market of activity trackers. First, this graphical overview reveals that each company makes its own strategic choice on the design clock of innovation pacing. Some companies have chosen to introduce new products frequently (i.e. Polar and Garmin) and others have temporized their product introductions with a lower frequency of multiple years (i.e. BodyMedia and Philips). Second, the overall market pacing (visualised with the dark hatched columns) indicates a three-year design clock for new innovations. The highest number of products were introduced in the total market for activity trackers in the years 2005, 2008, and 2011. Third, it shows that the number of competitors in the market increased exponentially, schematically shown with the grey dotted curve in figure 5.6, resembling the emerging stage of the characteristic life cycle curve of the market life cycle. The market forecast, based on extending the curve, predicts an increasing number of competitors as the market develops. The roadmapping team expects that the industry's sales will further flourish. This was confirmed by quick Internet research, which revealed announcements of several new product introductions in the years to come.

TIME PACING STRATEGY

↘ Industry synergy

In some fast-paced industries, such as the semiconductor industry, innovative organisations throughout the whole value chain come together to create an international industry roadmap. They discuss the roadmap such as for instance the International Technology Roadmap for Semiconductors (ITRS) does[8]. Organisations, such as INTEL, ASML, and Samsung, joined this initiative to discuss technology advancements and challenges on an industry level and make it possible to align scarce resources. With mainly engineers at the table, the common ground on the time pacing strategy was found in Moore's law illustrated in figure 5.7. George Moore found that from 1970 to 2001 the number of transistors on microprocessor chips doubled every two years. The ITRS roadmapping consortium projected this double density pattern on the future timeline of their IC roadmap. Figure 5.8 shows this in the ITRS roadmap regarding DRAM-chips. Interestingly, the ITRS consortium agreed on this two-year pacing for several years until the roadmapping participants began to discuss the fundamental technological constraints of geometrical scaling. They expected that the light wavelength that manufactures the chip with machines of ASML would become equal to the nanometer limits of chip function miniaturization. This constraint was expected to be reached within the future of the roadmap's timeline. Then, in 2007, by agreement of the roadmapping participants, the ITRS decided to slow down the industry innovation clock and change the pacing to three years (see figure 5.8). This exemplifies that although a technical performance measurement is used for an initial pacing, the innovation clocks are still an outcome of a strategic decision and not a mathematic measure. The future cannot be measured, only agreed upon.

To decide on the time pacing strategy for the design clocks, roadmappers can consider both the competitive timing and the industry timing[5]. The design clocks enable the long-term innovation rhythm of new design efforts.

↖
Figure 5.6
Time pacing analysis of product introductions by competitors in the activity trackers market.

cc Eva Frese, Niels Corsten, Robin Kwa & Robert Stuursma, 2013. PHILIPS Active Life Project report for the Design Roadmapping Master Course, Faculty Industrial Design Engineering,
Delft University of Technology.

↗
Car desk of the Time machine car that could 'fly in the Back to the future' movie.

cc CNN still from the movie.

↘ Time Performance of product visions

In the film Back to the Future, Marty McFly and Doc Brown travel to the future of Wednesday, October 21, 2015. This particular future point in time marked the future context with many fantastic products, such as flying cars, self-tying sneakers, and hovering skateboards. Some of these have come close to today's reality and others appear to be far beyond. Among the product visions from this 1985 film that have become reality by 2015, are 3D movies watched with VR goggles, fingerprint biometrics technology, and fax machines[9].

There are also product visions that have not become reality, such as smart clothes from which we see only some emerging signs today, but not the imagined adjustable size, self-drying, and programmable properties that were part of the film. Also, the movie's 'Dust-repellent paper' is not something we work with nowadays. Then there is another category consisting of product visions that have almost become reality, and might appear in some form in the future, such as hover boards and flying cars. So far, we have seen prototypes of these product visions, such as the P4 flying car of Dutch Design Studio Spark, and the Hoover boards, that now can roll, but might fly in the future[9].

Similar to this way of measuring the time performance of future visions, we carried out a research on product visions of 'old' roadmaps. The research focused on whether product visions with a particular future time point are more likely to get launched. We measured the launch performance of the product visions expressed on the roadmaps[10].

We 'travelled back' to investigate roadmaps originated in the years 2002 and 2006. Ten roadmaps and its reporting documents were collected from a balanced selection of business units' development programs of a multinational company in lifestyle appliances. We investigated 98 product visions that had been mapped to a particular future point of time. This total collection covered design value enhancements, user-centred value creations and new value propositions[9].

The research results show a high launch performance realisation of mapped product visions (see figure 5.9) - 74 product visions were actually launched – the relatively rate is of successful launches is 67 percent. This indicates that product visions that are mapped to a particular future point in time have a high probability of actually getting launched. This unexpected high realisation of the product visions, suggests that these product visions are more plausible. Only 24 product visions (24%) appeared not to have been launched at all; these were all product visions involving incremental value enhancements. Also, the launches with the longest delays—plus 2 to 5 years—were such product visions with value enhancements[9].

↗
Figure 5.7
Moore's law,
past analysis and future projection of functions/chip.

→
Figure 5.8
Semiconductor Industry roadmap by the International Technology Roadmap for Semiconductors-ITRS

Industry Roadmap of DRAM chips

cc Paolo Garini
Chair of ITRS.

TIME PACING STRATEGY

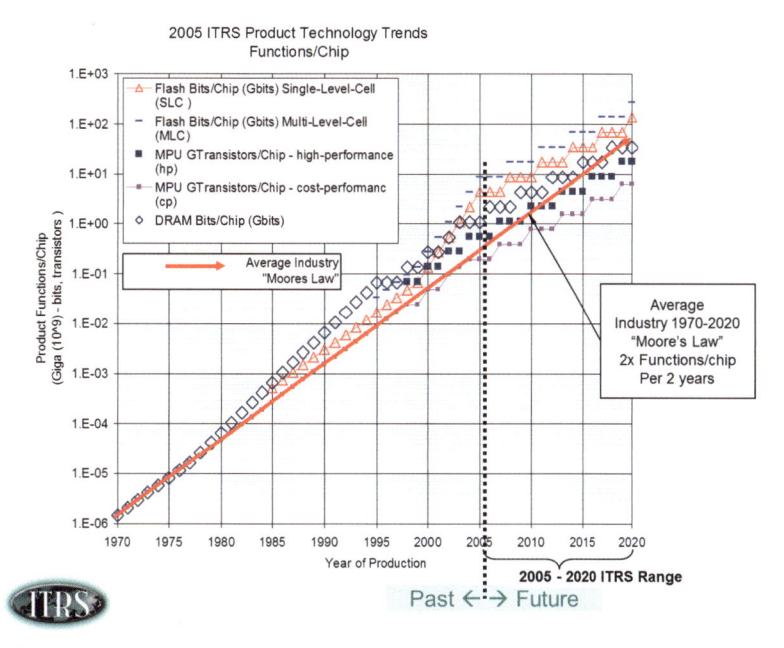

2005 ITRS Product Technology Trends Functions/Chip

Legend:
- Flash Bits/Chip (Gbits) Single-Level-Cell (SLC)
- Flash Bits/Chip (Gbits) Multi-Level-Cell (MLC)
- MPU GTransistors/Chip - high-performance (hp)
- MPU GTransistors/Chip - cost-performanc (cp)
- DRAM Bits/Chip (Gbits)
- Average Industry "Moores Law"

Average Industry 1970-2020 "Moore's Law" 2x Functions/chip Per 2 years

2005 - 2020 ITRS Range
Past ← → Future

INDUSTRY ROADMAP

2007 ('07-'22) ITRS Technology Trends DRAM M1 Half-Pitch : 3-year cycle   Update

Year of Production	2000 [Actual]	2001	2002 [Actual]	2003	2004	2005	2006	2007	2008	2009	2010	2012	2013	2015	2016	2018	2019	2020	2022
Technology - Contacted M1 H-P (nm)	180	151	130	107	90	80	71	65	57	50	45		32		22		16	14	11

2-Year Technology Cycle ['98-'04] → 3-Year Technology Cycle

DRAM TECHNOLOGY YEAR OF PRODUCTION	2015	2017	2019	2021	2024	2027	2030
Half Pitch (Calculated Half pitch) (nm)	24	20	17	14	11	8.4	7.7
DRAM cell size (μm^2)	0.00346	0.00240	0.00116	0.00078	0.00048	0.00028	0.00024
DRAM cell FET structure	RCAT+Fin	RCAT+Fin	VCT	VCT	VCT	VCT	VCT
Cell Size Factor: a	6	6	4	4	4	4	4
Array Area Efficiency	0.55	0.55	0.5	0.5	0.5	0.5	0.5
V_{int} (support FET voltage) [V]	1.1	1.1	1.1	1.1	0.95	0.95	0.95
Support min. V_{tn} (25C, $G_{m,max}$, V_d=55mV)	0.40	0.40	0.40	0.40	0.37	0.37	0.37
Minimum DRAM retention time (ms)	64	64	64	64	64	64	64
DRAM soft error rate (fits)	1000	1000	1000	1000	1000	1000	1000
Gb/1chip target	8G	8G	16G	16G	32G	32G	32G

We also measured the accuracy of the launch timing of the product visions (figure 5.10). These research results were less impressive. Only 28% of the product visions mapped to the timeline were launched on time. With a bandwidth of plus or minus one year, the user-centred value creations have the highest accuracy (62 %) and not the low risk visions of design value enhancements as could have been reasonably expected (46% time accuracy). The more disruptive new value propositions creation visions have a time accuracy of 27%[9].

Despite the seemingly impossible task of measuring future time in length and frequency (chronos time), the Back to the Future movie gave us the inspiration for this research. When you have an old roadmap and a future time scope within reach, we encourage you to do a similar performance measurement. To learn about the time journey a bit like the one undertaken by the teenager Marty McFly.

TIME PACING STRATEGY

↙
Back to the Future Car
1981 DMC DeLorean Time Machine, replica of the flying car.

cc Jack Snell.

↓
Flying car, PAL-V
(Concept) flying car designed by SPARK DESIGN.

cc Robert Barnhoorn.
SPARK DESIGN, Rotterdam, The Netherlands.

PHILIPS CASE

Wake up Light Roadmap

The PHILIPS Wake-up Light simulates a sunrise with a gradual increase in luminosity at the speed of a sun rise. This product is designed to make users feel more energetic and refreshed as they face the day ahead. Since its successful launch in 2006, five next generations have been designed around this user experience, and it is now used by millions of people in homes all over the world. This story is about the roadmapping project that I carried out in the year of its first launch. It is a short story about a steppingstone towards the creation of a successful business with increasing sales growth. It all started with one product.

The initial product design (see figure 5.11) came from Taco, the lead engineer who had put a lot of attention into the proof-of-concept, the functions, and the testing of performance quality. Art, the product marketing manager, chose France as the test market and the immediate lessons learned from the consumer response, including the opinion leaders of the press, concerned the aesthetics of the product. For this reason, I invited the lead designer, Bart, from the design department of the PHILIPS Consumer lifestyle division, to join the roadmapping team. He started working on the roadmapping vision by doing design research along the lines of the research work of Taco and Art. To complete the roadmapping team on future foresight expertise, I asked more experts to join: two technology experts from the PHILIPS Lighting and Corporate Research lab, and one marketing expert from the Market Intelligence department. We all came together to co-create the roadmap in several one-day workshop sessions.

↘ Three Horizons roadmap sketched

↗
User Value Insight on waking up like in the summer.

→
Figure 5.11
Initial Product Design

cc Taco manager front end innovation
PHILIPS VITALIGHT

→
Figure 5.12
Second generation wake-up light

cc PHILIPS DESIGN

In the first one-day session, we shared the state of knowledge in research. Bart took the sunrise as an inspirational source, presenting loads of visuals on the meaning of sunrise in areas such as art and architecture. At that time, he had uncovered that users were becoming more focused on listening to nature's rhythms, and linking this back to their biological clock. Based on this, and in combination with the consumer value driver formulated by Art, we conceived the overarching vision for the roadmap: to support the biorhythms of end users. Art had the insight that people have a strong desire to focus on seasonal changes and live a healthier life. The consumer research also showed changes in peoples' attitudes toward their bedrooms. Although the interviewed users still thought of the room as a very personal and intimate space, the team saw a change toward fashion and more frequent decoration, and an upswing in the importance of including aesthetic objects like vases and art. After having heard the

TIME PACING STRATEGY

presentations of other experts, including future foresight concerning technological directions such as the prospects for the timing of white LED technology, we collected and clustered all ideas from the members of the roadmapping team that appeared through active listening to these multidisciplinary presentations. Based on this, our lead designer, Bart, sketched the three horizons roadmap (see figure 5.13).

In the first horizon of this design roadmap, the product's aesthetics are the focal point; simple shapes, real materials, and a hidden user interface and displays enhance design value in this horizon. In the second horizon, the different alternatives for new technologies, such as flexible LED surfaces, displays, and external controls, are sketched together with armature architectures. The third horizon connects to the corporate ambiance vision, focusing on the creation of end-user experiences in its context. In the bedroom context, we envision that the wake-up light will be connected to all the lights in the room to create an immersive sunrise experience.

The transition from the first to the third horizon includes changes from a pure, standalone design of a bedside lamp into a connected system of personally combined lamps that are universally controlled by the sunrise controller that we patented.

↘ Design clocks

As a result of the next roadmapping sessions, the product marketing manager, Art, created the product roadmap. We discussed the time pacing of the emerging business of the wake-up light and consulted Art's colleagues in more mature markets about this issue.

For reference purposes, we adopted the three-year design clock for next generations that Philips uses in its similar lifestyle businesses.

Furthermore, we decided together with the team to pinpoint the first horizon vision on 2008, the second on the next two to three years

(2010/2011) and the third four years after the first horizon, from 2012 onwards.

For the innovation clock of a new model that enhances the first wake-up light, we had estimated that it would take a two-year innovation effort to enhance the value of the wake-up light by creating a more compact and contemporary aesthetic for it. The three-year projections for the second horizon product visions concern new user-centred technology applications, such as market testing sales of a standalone controller called 'distributed light'. The longest-term innovation that we envisioned is an open-system innovation for an immersive sunrise experience in the bedroom.

↘ Time performance

The time performance of this design roadmap— in the sense of looking back to the future and measuring the launch performance of the product visions with a particular future time point —appeared to be surprisingly successful.

In 2009, one year later than projected in the roadmap, the second generation of wake-up lights was launched on the market (see figure 5.12). This new generation gained outstanding design recognition, with both the iF and red dot design awards. The design achievements in the leap from the first to the second generation include a simplified product design that removes the base section, effectively turning the entire object into a light. Also, a sound artist created a better range of noises for the alarm. Overall, by moving the display panel inside the product, this next generation product offers design value that both consumers and the industry loved.

Later, the user-centred innovations introduced in this successful generation were enlivened further in the third and the fourth generations, which were launched in 2011 and 2014, respectively. The time performance of these generations is also within the span of one year. Both generations again had a more functional focus, this time on energy savings and iPhone integration. For the realisation of the third horizon vision of the original roadmap, there are still design challenges to tackle. In the meantime, the original roadmap has of course also been adjusted and adapted to the new design research findings.

DESIGN ROADMAP

Wake-up-light
PHILIPS

→
Figure 5.13
Sketched Design Roadmap for the Wake-up-light

© Bart Massee, 2006
PHILIPS DESIGN.

167 TIME PACING STRATEGY

SURROUND SUN SYSTEM

M3

INTERIOR

ENVIRONMENT

GLOBE

FLEXIBLE

TOWARDS LIGHTING

LEDS SURFACES

REMOTE

6.45 WAKE UP

SOFT FABRICS

GLASS

SNOOZE

EXTERNAL CONTROLS

Wake-up Light RDMAP

Philips / Massee 07/06

LAB ↗

Map design clocks

MATERIALS NEEDED:
- → note book
- → laptop
- → large sheets of paper
- → sticky notes
- → pens

1 A good way to start the timeline design of your roadmap is with a historical timeline. Sketch or print out a large one and put it up in your workspace.

2 Now mark the key moments of prior product/service launches of the business you are working for or, if your business is just starting up, for a similar business category.

3 Answer questions like, What is the time interval between two incremental innovations of design value enhancement? What is the innovation clock for these model changes? What is the time interval between two new user-centred products/service designs and between radically new value proposition designs? Take these intervals as the starting points for your design clocks.

4 Now that you've got a sense of your design clocks, look at the past launches of one to three competitors. For instance, you may choose to study the market leader, an innovative new entrant, and a similar size company. Carry out the same retrospective analysis on the time intervals for the different types of design innovations. Fine-tune your design clocks with this market information.

TIME PACING STRATEGY

5 Consider if the users you work for have expectations regarding critical new technologies. If they do, look at the suppliers of these technologies and analyse the innovation clock of subsequent technologies. Fine-tune your design clocks with this technology information.

6 Now sketch or print out the future timeline for the roadmap and put it up in your workspace. Label the start point of Horizon 1 on the timeline with the present year (and quarter if you like). Build it up with three horizon sections. For the endpoint, roughly indicate a year in which the radical new value proposition design, of Horizon 3, will be launched. For Horizon 2 indicate a year in between for user-centred design.

7 Decide on sensible time intervals for the design clocks of your future launches. Mark these key moments of first exposure of the designed solutions to the users on the future timeline. Consider the key moments for the different types of innovation, value enhancements, value creation, and value proposition development. The number of key moments you map may vary. Consider what might be most critical to the persons for whom you are designing.

8 Mark the key transition moments. They could be seasonal changes, important expos, public holiday dates, or recurring deadlines of budget reporting.

As you set out to design a roadmap, your first mapping challenges are the timeline and setting the speed of innovations by the creations of design clocks. This gives you a chance to do some research on the timing of the business you work for. Reflect on the historical timing of valuable innovations launched by the business, its competitors, and its partners. Getting a good handle on the design clock can set a structure and rhythm for the temporary activities of design innovation.

You can use this first mapping of design clocks as a starting point for building a more descriptive roadmap. Your roadmap will change as things evolve, and that's perfectly okay. You can always amend things.

Prof. dr. ERIK-JAN HULTINK on the subject of Launch Strategies and Roadmapping

LS Your product launch research shows that to time a launch successfully, it is important to make trade-off decisions with new product sales income[11]. What is the important takeaway here for roadmappers?

EJ The funny thing about the results of our research is, that although commonly people take into account the revenue for decisions on the timing of new product introduction, we looked at companies' actual performance data and found that for the optimal timing of new product introductions - the more relative account of both cost and revenue give the best chance of new product success. So, contrary to the common belief that speeding up your product innovations is always the best thing to do, we found that speeding up most of the times also comes with much more costs. And these cost need to be balanced with the expected revenue estimations. Hence we suggest to decide on the optimal timing of new product introductions, in the sense of a time pacing strategy. Our research also showed the differences - between really new products, improvements and line additions. Basically we evidenced that a too early launch timing can be equally bad as too late. Therefore, I would like to recommend that for any time pacing decision, you take into account both the cost and sales consequences of your timing .

LS In roadmapping, product visions are mapped on a timeline. How do we get the entry timing 'right' in relation to the new product's window of opportunity?

EJ David Bowie released Space Oddity, one of his most famous songs, just ten days before the first man landed on the moon. And his song was actually used in the BBC's coverage of that unprecedented event. This was a matter of timing and opportunity. Basically, you do not necessarily have to be the first, but you shouldn't be seventeenth either. So there is this 'window of opportunity' in which you still can be in time, when the market is growing, and people are talking about it and are open to it and you also have access to the retail shelves. If you wait too long the window can be closed. If you launch too early, the window might not be open yet. But this does not necessarily mean that you always have to be the first. Take for instance the electric car. How many decennia have we been yakking about it?

LS   Well, hybrids were introduced first.

EJ   Launching the hybrid was probably needed to improve the chances that electric cars would be accepted. I believe Renault was the first to introduce electric cars way back in the 1930s, and in the 90s they launched the Clio Electrique. It was hugely expensive and only ran for an hour and a half off one charge. Back then, the timing wasn't right – the window of opportunity wasn't open yet. But I just recently saw an amazing video about – on why we all will drive an electrical car within the coming 30 years. In this video a traditional car and an electrical car were tear down in parts. . The screening of the traditional car — think about a BMW or the like- showed about 500 parts and the Electrical car — Tesla S or similar - showcased 20 parts. In one instance you grasp - without the 'proclamation' stories about saving the planet - that on the long run it will never be possible to be cost-efficient. You can never reduce the traditional system cost until the level of the more optimal system with also the additional benefits of easy of spare part replacement. It just cannot be. This example also showcases that it is much easier to do time pacing with less parts. When your product has a modular structure, it is easier to upgrade your product line, and it is also much easier to organise when you have 20 parts instead of 500+.

LS   How do you explain the fact that so many roadmaps with product visions – 74 out of the 98 we studied[9]– successfully led to real products?

EJ   I assume that working with visions and making them explicit, generates the management commitment to realize it. When you map the new product visions and eventually allocate people and money to it, it is not a 'free-floating' idea anymore. It becomes real, projects get started with a roadmap. When I consider that, over the 20 years that I have been doing research on innovation at organisations, I have never come across a situation with extremely few ideas. Always there were too many ideas for too little time and

Figure 5.9

Launch performance of the product visions

Figure 5.10

Time performance accurancy

cc Simonse & Hultink, 2016[9].

too little money. Roadmapping supports the making of choices and priorities, in a way it makes an idea strategic – in the spotlight for the whole organisation, and 'making the idea happen' becomes serious business.

LS   What is your explanation for the low time accuracy performance we found on the product visions versus the actual product launch timing[9]?

EJ   Time-to-take-off for really new products almost always takes longer than estimated in advance. Often that is due to the fact that the introduction stage of the product life cycle is often a long uncertain stage. And from other research we know that radical innovations take more time to introduction due to more uncertainties. Incremental enhancements are variants on a known path. You have the overview of the steps you have to take, so you can estimate this more accurately. Remember 1999, the Prince song? -Tonight we are going to party as if it is 1999 - It came out in the early 80s.

LS   Interesting music metaphor! Here's one more question: what do you recommend design roadmappers to do in creating a time pacing strategy?

EJ   Make it visual. Do not only use text and numbers in a roadmap. In my experience from strategic workshops, one activity that always works out well is visualising the connections between the innovation project, or projects, and the innovation strategy. For this visualisation activity we used one set of cards to draw and write the projects on, and another set to visualise and describe the focus elements of the innovation strategy. This is often a revelatory experience for the managers involved – they actually see links missing, and there is often an overload of cards with cost reduction projects alongside only a few cards representing the heart of the strategy. Visual mapping really opens eyes.

ERIK JAN HULTINK is a Professor of New Product Marketing at the Faculty of Industrial Design Engineering, Delft University of Technology, The Netherlands. His research focuses on launch and branding strategies for new products. He has published on these topics in such journals as the Journal of the Academy in Marketing Science, and the Journal of Product Innovation Management. He was ranked number six in the list of the World's Top Innovation Management Scholars, and selected as the most productive European researcher publishing in the Journal of Product Innovation Management. He is co-founder and board member of the Dutch chapter of the Product Development and Management Association (PDMA). He regularly consults companies on the topic of new product launch, and frequently appears on the Dutch television and radio commenting on the success and failure chances of new products.

TIME PACING STRATEGY

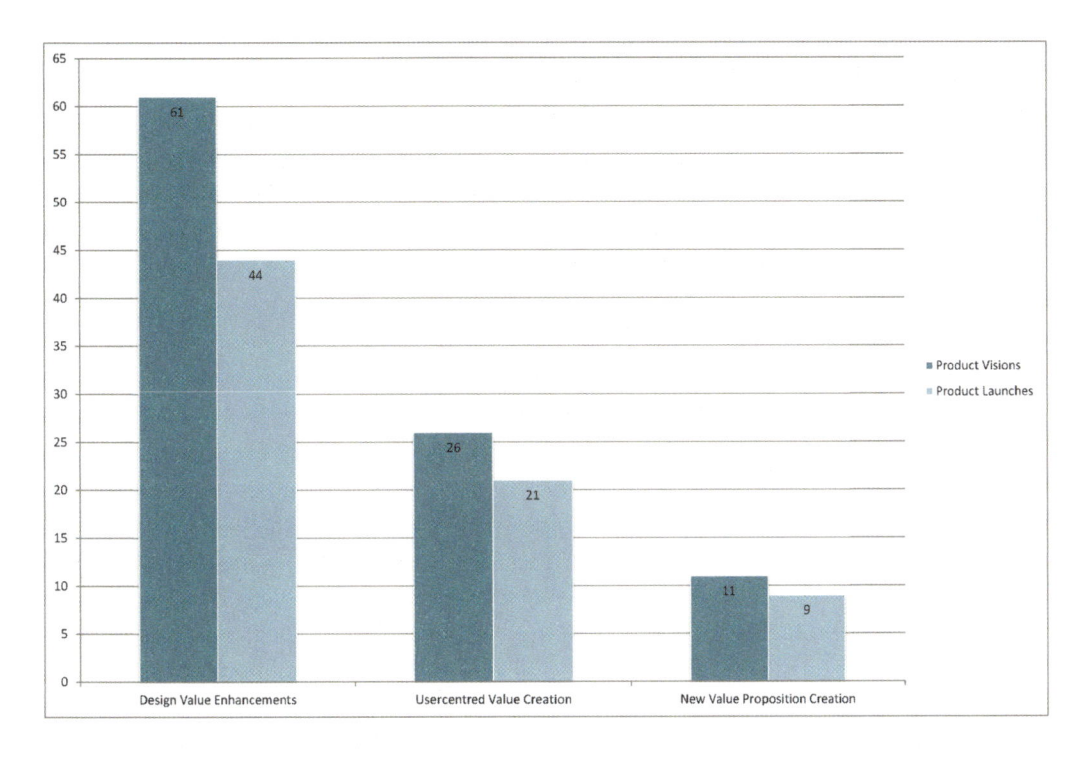

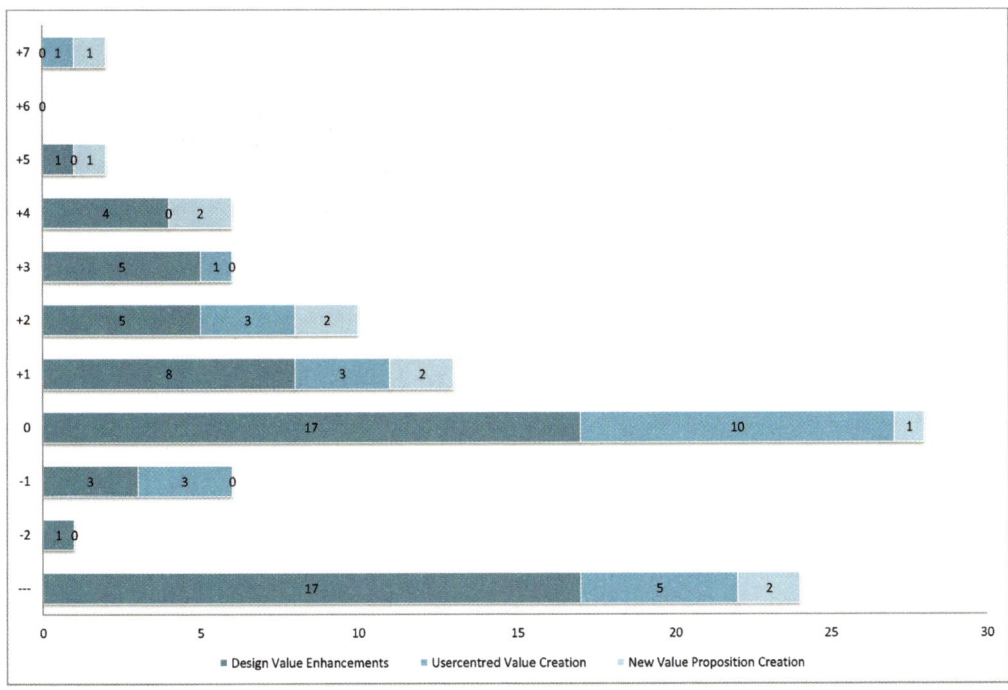

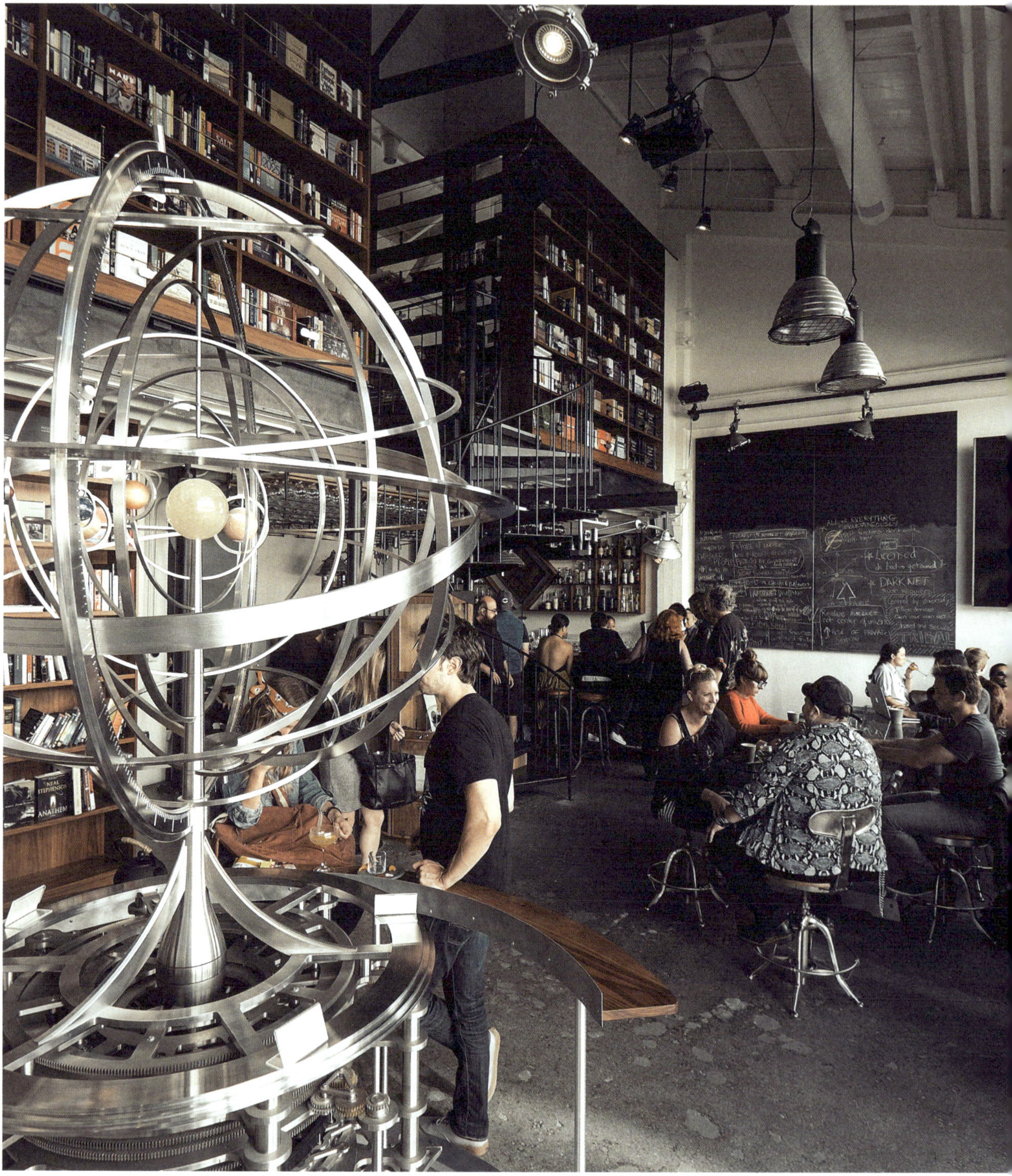
NEAL
STEPHENSON
ANATHEM
DARK NET

IN SUM

Coming back to kairos and perceiving time as opportune time, the Three Horizons concept offers such a mode of thinking about the future. The three horizons describe three different spaces of opportunity moments and focuses the attention on three parallel strategic life cycle evolutions. Furthermore, the key in the answer to the questions of how to decide on the timeline design is in the design clocks.

Design clocks include three types of pacing for different types of design innovation. Model changes have a cycle time of one year or two years, while platform innovation with new user value creation has a cycle that is two or three times longer. New business development by means of value proposition creation takes the longest, with a pacing time interval of six to fifteen years, depending on the capability complexity of the industry.

The pacing of design clocks in roadmaps is decided upon by user-driven target group expectations concerning model year changes, by technology-driven performance measurement indicators, or by design-driven value enhancements, depending on the type of design roadmap in question. In industry roadmaps, the engineers around the table prefer to pace with an exact technical performance measurement. In company roadmaps in the consumer electronics and appliance industry, the pacing decision is also based upon the pacing of competitors. In design roadmaps, these two tensions are bridged by user-centred pacing driven by target group expectations. Design-driven roadmapping thrives on value creation.

The guidelines of this chapter include the time-based framework of design innovation that frames the three design clocks with three horizons of evolutionary design innovation. Its use was showcased with the Philips wake-up light roadmap, also in a visual way by drawing the three horizons. The lab provides a short guide to experience the art of time pacing.

←
The Long Now
Clock prototype, the interval.

© Christopher Prentiss Michel, photography.

The Long Now foundation hopes to provide a counterpoint to today's accelerating culture and help make long-term thinking more common.

---

1. Von Leyden, W. (1964). Time, Number, and Eternity in Plato and Aristotle. The Philosophical Quarterly (1950-), 14(54), 35-52.
2. Curry, A. & Hodgson, A. (2008). Seeing in multiple horizons: connecting futures to strategy. Journal of Futures Studies, 13(1), 1-20.
3. Simonse, L.W.L. & Hultink, E.J. (2017). Design roadmapping: Managing transitions of the strategic life cycles. 24th Innovation and Product Development Management Conference (IPDMC), Reykjavik, Iceland, 11-13 Jun 2017.
4. Willyard, C.H. & McClees, C.W. (1987). Motorola's technology roadmap process. Research Management, 30(5),13-19.
5. Simonse, L.W.L., Hultink, E.J. & Buijs, J.A. (2015). Innovation roadmapping: Building concepts from practitioners' insights. Journal of Product Innovation Management, 32(6), 904-924.
6. Bayus, B.L. (1994). Are product life cycles really getting shorter? Journal of Product Innovation Management, 11(4), 300-308.
7. Eisenhardt K.M. & Brown, S.L. (1998). Time pacing: competing in markets that won't stand still. Harvard Business Review, 76(2):59-69.
8. Lange, K., Müller-Seitz, G., Sydow, J. & Windeler, A. (2013). Financing innovations in uncertain networks—Filling in roadmap gaps in the semiconductor industry. Research Policy, 42(3), 647-661.
9. Simonse, L.W.L. & Hultink, E.J. (2016). Back to the future: Product visions and product launches. 23rd Innovation and Product Development Management Conference (IPDMC), Glasgow, United Kingdom, 12-14 Jun 2016.
10. Hultink, E.J. & Robben, H.S. (1995). Measuring new product success: The difference that time perspective makes. Journal of Product Innovation Management, 12(5), 392-405.
11. Langerak, F., Hultink, E.J. & Griffin, A. (2008). Exploring mediating and moderating influences on the links among cycle time, proficiency in entry timing and new product profitability. Journal of Product Innovation Management, 25(4), 370-385.

MAPPING SESSIONS

HOW TO ORGANISE THE ROADMAPPING PROCESS

Elementary in the process of roadmapping is to create a mutual understanding of the promising opportunities, the current positioning, the future vision and the pathways towards it. In order to have constructive conversations, you can arrange a set of mapping session for your roadmapping team. How to organise this is the focal topic of this chapter. So far, we have addressed the diverging activities of creative trend analysis and technology scouting, and the converging activity of future visioning and time pacing. In this chapter, we address the challenge of bringing these together in mapping sessions.

Mapping sessions enable the dialogue through the creative facilitation with maps, the most promising innovation options are discussed, mapped and synchronised across-functions. An experienced roadmapper at the gasoline company BP expressed once: "The structured dialogue is essential to the foresight process; not be unduly sensitive to the opinions of individual gurus, nor be over-reliant on existing organisational structures and power bases, ensure discussions are informed, open, and objective to help greatly in making this consensus-building process both efficient and effective[1]." Underlying the roadmapping process of diverging and converging activities lays a process of consensus building. A process that is sensitive to personal characters, styles of communication, attitudes and behaviour but also to structure, creative facilitation and outcome. For your support, this chapter provides you with insights and guidelines on the following topics:

- → Organise creative dialogues;
- → Facilitate the mapping in three critical sessions;
- → Support strategic decision making.

The organisational challenge is to orchestrate a seamless chain of mapping sessions.

Creative dialogues

↘ Dialogues

Many roadmappers subscribed that one of the greatest values of roadmapping lies in the process of communication. The crucial importance of using dialogues to talk about the future and share imaginations, gut feelings and beliefs emerged from our research on 12 cases of roadmapping[1]. The majority, 9 out of 12 firms, explicitly mention the importance of dialogues in roadmapping. Firms such as Motorola, Philips, Lucent, Tata steel, Sandia, ABB and Lucas Varity reported that structured dialogues are part of the consensus-building process. As a senior roadmapper at Tata steel –Hoogovens expressed: "A unified approach ... promoted growing commitment to the common decision[1]", and another roadmapper at Sandia, explained the process of roadmapping as to: "Develop a consensus about a

179   MAPPING SESSIONS

CREATIVE GROUP CONVERSATION about the future plans on innovation with the timeline as a focal point for creating mutual understanding.

set of needs and the technologies required to satisfy those needs[1]."
Several roadmapping experts emphasized this underlying process of consensus building. To fully grasp what that is, the legendary professor Schein, explains on dialogues of consensus-building, that they draw on

"a conversation that deliberates basic choice points, with room for personal evaluation of options, and with taking time for the suspension of disbelief, making a collective choice by internal listening and accepting differences and building mutual trust[2]."

Organising these dialogues includes the creative facilitation with maps for visual interaction and group conversation that are aimed at the 'suspension of disbelief'.

↘ To synchronise

Roadmapping session thrive on creative dialogues in which the team creates a collective understanding of future innovation plans. Members from design, marketing, research and development come together to create the sense of urgency around innovation deadlines for the future[3]. As explained by an experienced roadmapper, a mapping session: "places greater emphasis on, and provides practical assistance to facilitate the construction of the innovation plan by enabling all the participants, whether technically expert or otherwise, to play a full part in the process[1]."

A creative dialogue stimulates the cross-functional communication in which the use of visual maps instead of only textual material is vital[1]. The maps with the future timeline on it support the synchronising of directions on innovation efforts. During the creative group conversation you can synchronise the elements of the future plans on innovation with the timeline as a focal point for creating mutual understanding. The facilitation of the dialogue concentrates on having the individual and team energy around the common goal of the future vision. Typically, the group gradually builds a shared set of meanings about the vision on future innovations related to a certain point in time.

↘ Maps for visual interaction

Unlike 'discussion' and 'debate', creative conversations not only cover speech but include a rich mix of visual maps. The visual mapping of the contributions of the participants is crucial. In essence, roadmaps are created out of the visual interaction with the maps. According to a roadmapping expert of Tata Steel the maps "promoted growing commitment to the common decision[1]." The maps help to focuses the team's attention on creating future directions of promising innovations. Depicting the interrelationships and structuring the sequence between ideas and activities resemble the innovation paths of the future.

Overall, roadmappers have experienced that "The use of visual outputs has been found to help greatly in making this consensus-building process both efficient and effective-BP[1]." For structuring the dialogue, you can use different types of creative facilitation techniques: ranging from common brainstorming, gamestorming and idea generation techniques, to future dedicated Delphi techniques and scenario tools. Several dedicated books and sources on the Internet can provide you with guidelines on these. To get started with facilitating the mapping sessions in a creative way, there are two initial techniques worth considering: affinity mapping and matrix mapping.

↘ Affinity mapping

The technique of affinity mapping organises a large number of facts or ideas into their natural relationships. This method taps a team's analytical thinking as well as creativity and intuition. It was created in the 1960s by Japanese anthropologist Jiro Kawakita[4].

Designers typically use affinity maps when they are confronted with many facts or ideas in complex contexts when issues seem too large to grasp and when group consensus is necessary. Therefore affinity mapping is very suitable for a roadmapping session. It supports in understanding the complex environment of an organisation and to grasp the values driving the future innovations.

In a mapping session with presentations (on for instance creative trends or scouted technology ideas) you can ask all members to listen critically to the presentations, and actively map their thinking on sticky notes. You can ask for instance to actively listen for three things: values, opportunities or problem issues.

In between presentations the sticky notes can be put on a poster with a draft roadmap on the wall. By affinity mapping, you can organise the sticky notes into logical categories and groups, and see the themes appear visually: see the values, challenges, and opportunities. Grouping the problems together also provides an

important base for triggering innovative thinking.
    To wrap up, facilitate a round of collective sense-making to see the big patterns and themes – drawing insights by defining common themes and attributes and identifying the tension points.

↘   Matrix mapping

The matrix technique positions, seeks and structures the relations in a two-dimensional way. This technique supports a team's cross-functional thinking on a strategic level with 2x2 matrices. The Dutch roadmapper Groenveldt invented the innovation matrix as a base for roadmapping in 1986 and since then many variations on the matrix technique followed[5].
    Designers typically use matrix maps when they want to position their design innovations with current - and new user target group, current - and new technologies. In a team setting, design roadmappers make use of a matrix to map interrelations between values, ideas, technologies, products, functions, features, trends, etc. The matrix zooms in on the relations between two layers of a roadmap. Such as the matrix map with the value tree of a user target group on the x-axis and the propositions features on the y-axis. Or a Functions-Technologies matrix, more commonly known among engineers, with on the x-axis the functions of the product line releases and on the y-axis technologies[6].
    For the matrix you select two innovation dimensions and prepare the matrix map for a session in which you preferably invite equal as much professionals with knowledge on each of the two dimensions. By a pitch or short introduction of the items on the matrix, a discussion follows on its potential for connecting to an item on the other matrix dimension. More options or no options are also possible outcomes of this discussion.
    Matrix mapping creates a more detailed view on the interrelation of the promising options for the future. Many roadmappers have designed their dedicated matrices map for facilitating a mapping session on a particular strategic issue concerning the roadmap.

Three critical mapping sessions

You can organise a design roadmapping process as depicted in figure 1.3 in the first chapter, with a chain of mapping sessions. To provide a minimal critical specification of a roadmapping process, we picked out three sessions. For each stage, one session: (1) the value mapping session, (2) the idea mapping session and (3) the pathway mapping session. These sessions are minimal needed for sharing the results of the main roadmapping activities. The first session is concentrated on the

Three critical MAPPING SESSIONS are (1) Value Mapping (2) Idea Mapping and (3) Pathway Mapping.

↑ ↗ →
Creative conversations

© SUNIDEE photography

Workshop with Unilever innovation professionals.

MAPPING SESSIONS

mapping of user values, the second on generating ideas in connection to technology modules and the third session goes into the details of programming design innovation projects. The three critical sessions, lead respectively to the deliverables of a future vision, a design roadmap and a design program roadmap

↘ Value mapping session(s)

Part of creating a future vision is to organise one or more value mapping sessions or workshops. To prepare for these creative dialogues, you decide whom to invite among the innovation professionals. Besides the roadmapping team members and those who have been active in the roadmapping activities of creative trends research and technology scouting, we recommend you to invite senior managers and whenever appropriate business to business clients, strategic partners and preferred suppliers.

Compose the invitation list for the session in such a way that you ensure a solid base for creating common ground on the formulation of the future vision. For the imaginative inspiration, the invitation of 'mavericks', and 'rebels' with strong stories on the future, is also worthwhile. When your list of invitees reaches 25 or more members consider to organise more sessions. You can do this in a parallel way for instance by organising a large scale future search conference, or in a sequential way by arranging a set of sessions with different participants in addition to a core team of roadmappers. You can also consider involving a facilitator or dedicated consultant for organising the future vision session(s).

Take as time reservation at minimum a four hours for a 'pressure cooking' workshop, and at maximum a three days future search-conference session .

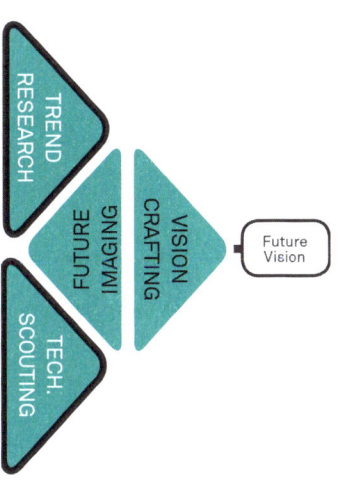

This value mapping sessions brings together the results of three roadmapping activities, of creative trend research, future imagening and technology scouting (see figure 6.1). The purpose of this session is to craft a future vision that draws on the outcomes of these activities. Those who accumulated knowledge on trends and technologies can kick of the session with sharing their findings and insights. The other participants are asked to 'dump' their ideas on user values on sticky notes while listening to the presentations. With affinity mapping, you can capture many fruitful insights on values that are triggered by the sharing of the state of knowledge as it is

↗
Figure 6.1
Activities of Value mapping

cc Simonse, 2017.

today.

The creative conversation further concentrates on mapping these fresh ideas on values. Then a sensitizing activity in subgroups can fire start future imaging. You ask the participants to create vision statements on the 'Desirable future of our organisation',(See the Lab exercise). You capture, cluster and rephrase the values that are important for the future. After this immersing in a future context, the decisions to be made in this session are about the vision direction.

The value mapping for this starts with creating an 'agreed on' list and a 'disagreed on' list. It is important to reach consensus on the future vision that can meet the properties of future visions: value drivers, clarity and magnetism (see the third chapter). You might want to support the value mapping by sketching the future images that are put forward by the participants and build up a raw version of the three horizons roadmap as shown at the wake-up light case (figure 5.13).

More than one session might be required as a follow-up to generate clarity and consensus about the most important user values for the future vision. As a follow up you can visualise the future vision and shape it into an artist impression with the vision statement mapped as direction for the draft roadmap.

Idea mapping session(s)

At the heart of roadmapping is the idea mapping on user values in association to technology modules. You can invite the same participants as those in the value mapping session or change the extended team composition in such a way that you also involve the people who are good in generating ideas for new products and services. Make sure that you also invite key persons that represent the different domains of technology, marketing and design. Together you will work on generating new concepts in this session.

The purpose of this session is to discover matches of user values and technologies that can also provide interesting opportunities for the future business development. As shown in figure 6.2. the three activities that are integrated in this mapping session are: the time pacing strategy, the user group value insights and the technology module options. As backbone for the generation of new ideas you can use the time pacing strategy mapped on the timeline of a draft roadmap. (see the Lab, in the fourth chapter Time pacing strategy, on how to create one).

The creative lead can kick off the session

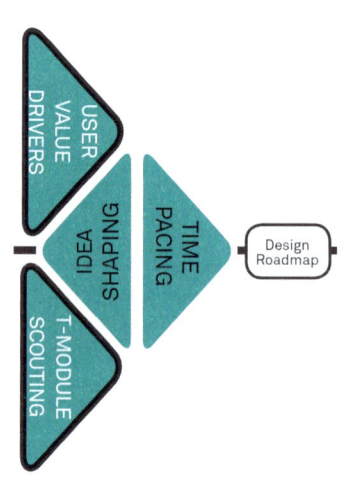

Figure 6.2
Activities of Idea mapping

cc Simonse, 2017.

MAPPING SESSIONS

↑ ↗ →
Creative conversations

© SUNIDEE photography

Workshop with Unilever innovation professionals.

with this backbone of the time pacing strategy, by pinpointing the 'blank spots' of line extensions, new platform, and new value proposition innovations. The team members who have carried out design research on the product service architecture can bring in their findings and together you map the first recombination's on the draft roadmap.

For this session you can consider to use the matrix mapping technique with the value tree maps on one axis representing important user values and promising technologies harvested from the technology scouting as the other axis. Then in subgroups the participant can work on generating new ideas in connection to a specific user value, potential technology option. With a chosen technique of for instance brainstorming or gamestorming, the sub-team members can build upon each other's ideas about future products and services. The ideas are mapped, grouped and ordered around innovation themes and prioritized blank innovation spots. In a second round of idea mapping the sub-teams are encouraged to further clarify the meanings and back ground of one particular theme of future options they want to bring to the table. Depending on the time you reserved you can organise the agenda for this session in such a way that subgroups continue with a second round of concept enrichment in relation to certain parts of the roadmap and have a collective presentation of the outcomes, or you choose to organise successive ideation sessions on future concepts.

The decisions to be made in this session are about the concept vision of the product/service. The concepts with the highest impact on user value and the best fit with the innovation capabilities can be selected for the blanc options of the time pacing strategy. Consider to map one or two alternatives per strategic option on the draft roadmap, to allow for a fall back scenario.

As a follow up you can visualise the product/service roadmap and shape layers for the technology roadmap and the user value roadmap (see the template of figure 7.3 in the next chapter). You can stick it on a wall, for communicating about it in an informal way not only with those who have participated in this session but also with the other stakeholder audiences. We have experienced great benefits of doing this in a dedicated room the so-called 'future lab for innovation'.

Pathway mapping session(s)

In this mapping session, the innovation program for the mid and long-term is created and this explicitly involves the allocation of resources and creation of deliberate order in a pathway. This mapping converges the results of all previous steps in a draft design program roadmap through the parallel alignment of the design roadmap and the program roadmap.

In this session you create the pathways of innovation and depict the relationships between the innovation elements. Pathway mapping concerns activities of creating pathways by allocating, ordering, and

MAPPING SESSIONS

↑ ↗ →
Pathway mapping

© SUNIDEE photography

Workshop with Unilever
innovation professionals.

→
Figure 6.3
Activities of Pathway mapping

cc Simonse, 2017.

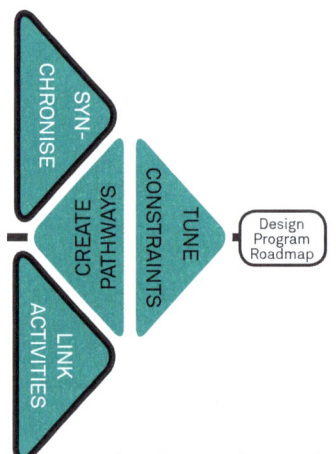

linking innovation activities of technology foresight, long-term market encounters, and product line evolutions on a map with a timeline related to the future. It concerns the synchronising of multiple innovation elements and the creation of linkages between these innovation elements. (see figure 6.3). The purpose is to map the activities and its resourcing within the constraints of the resource limits. Ideally, the cross-domain relations are discussed in a 1-2 day workshop with a multidisciplinary team of members that represent the innovation domains.

Invite the roadmapping team members and those who joined the previous sessions, including the creator of the product/service roadmap and key projects programmers across domains. At the beginning of the session, brief the team on what the roadmap should communicate to whom (see the next chapter about target audiences for roadmap communication).

Have a schematic overlay of the roadmap template you prepared available in a huge version on the wall, and distribute a big pile of empty cards. Together with the team, identify the project activities for the roadmap: one item per card. Next, identify relationships and fine-tune content (iterative process).

The first 'fine tuning' round draws from the product roadmap. You can create pathways of innovation by mapping the evolution of product lines in small steps of logical updates. Then relate the user value and the technology module to each of these steps and draw this relation on the roadmap.

The second 'fine tuning' round brings in the resource constraints. The financial constraint of the total innovation budget and the capability constraint of the total work force on innovation. Also certain capability limitations of specific expertise need to be addressed here. In this fine-tuning activity you translate the innovation pathways into project activities. You aim for creating a program roadmap that can be used, and updated in strategic reviews and alignment meetings with partners and suppliers.

There are many detailed decisions in this session about the alignment of relations. The timeline supports the alignment in discussing the synchronisation towards innovation targets of for instance launching a product. The key decisions concern the strategic agreement on the new products and services for the future and the projects to initiate for the first upcoming year.

A follow-up activity of this session is to compose the project briefs with elements from the roadmap. Other follow-up activities include a

MAPPING SESSIONS

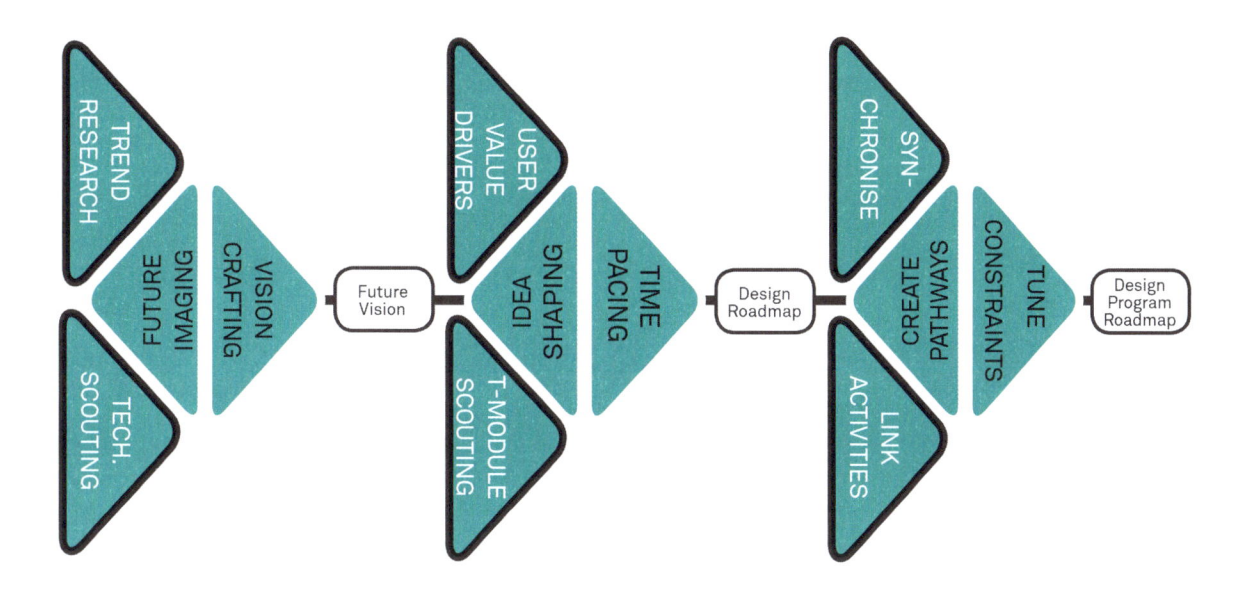

budget and resource planning validation. The final roadmap outcome provides the outline for project proposals of the future. In addition to the baseline overview of the roadmapping proces (figure 1.3), above figure 6.4 presents a detailed proces overview in relation to the critical mapping sessions.

Overall, the mapping sessions are about seeking to achieve a common view, developing consensus, and obtaining commitment to a unified innovation plan[7]. Synchronising around the timeline appeared to be crucial for the consensus-building process of communicating together and persuading stakeholders in the roadmapping discussions. The creative facilitation focuses the group's attention on structure, interrelations and sequences with respect to providing future directions of the strategic intent of innovation. The use of visual outputs has been found to help greatly in making this consensus-building process both efficient and effective[1].

↑
Figure 6.4
Detailed activities of Design Roadmapping proces in relation to the mapping sessions.

cc Simonse, 2017.

Decision making

↘ Strategic review: filtering potential options for the roadmap

Roadmapping involves making strategic choices. From all the spotted trends and technologies only those that offer real opportunities and are within reach of the available innovation capabilities are relevant. A common practice in a roadmapping session is therefore to review on strategic attractiveness and fit. In the particular context of strategic design innovations, the attractiveness is determined by the potential user value[8].

IS THE USER VALUE REAL?

Is the most important question for pursuing capability investments in design innovation. Numerous research studies on innovation performance have confirmed that new customer value propositions have only chance on success when the designed innovation solution can connect as closely as possible to the user's need or desire. Then, the second most important question in making strategic choices for innovation efforts is[8],

CAN WE DO IT?

Does our organisation have the capabilities to offer the desired solution? Or might we easily develop the newly required capability by leveraging on our strengths. Innovation success has been proven to be determined by the core capabilities of the organisation[8]. Therefore the two main criteria that enable the strategic design innovation are the fit with innovation capability and the impact on user value (see figure 6.5).

In a roadmapping session, this job of strategic review, can be done by the participants, who rank the innovation options (for instance the trends, or the technologies) in order of impact on the user value, impact from high to low. Followed by a second ranking on the strategic fit with the innovation capabilities. Plotted on this grid with two axes, the options with the most impact and most innovation capability fit, are selected as most interesting for building the roadmap.

↘ Decisions of Pursue, Park and Drop

With three classifications on user value impact: major, significant and low. (see figure 6.5) and three classifications on innovation capability fit: major, significant and low fit. The decision for options with low impact and low fit is to "drop from list". These options are dropped. Choosing

Creating a roadmap involves MAKING STRATEGIC CHOICES, and filtering out unviable options.

involves selection and filtering. Especially important in roadmapping that only those options that are strategic relevant are mapped. For choosing the design innovations with the highest potential these two criteria are used. Only high impact and high fit options are selected for mapping on the roadmap. For the others the decision is to drop it or park it on a list for periodic review.

Sometimes, for more robust decisions that also overcomes hesitations and problems with making a choice, after all dropping can sometimes be difficult, two additional review questions can be added that concern the market arena with the competitors and the internal expertise of the people working on innovation. The two questions are[8]:

> DOES THE DESIGN INNOVATION OFFER COMPETITIVE ADVANTAGE?
>
> CAN WE WIN?

The first question relates to the uniqueness of the value in the opportunity or generated concept. The second question relates to if you can meet the competition. When the value has clear advantages over alternatives, such as greater safety or social acceptability and when the organisation has experience and skill advantages in house, and these superior skill levels also match with the design innovation scale and complexity, in such a way that it can survive competitors' responses (such as price and patent war), the decision to pursue makes sense. When in doubt about this, the sensible direction of the decision is to park or drop it.

A last important question in strategic choice for a design innovation is[8]:

> IS IT WORTH IT?

Will it be profitable at an acceptable risk? When the forecasted returns are greater than costs and when the financial risk are acceptable – considering matter such as timing and amount of capital outlays, marketing expenses, breakeven time, and the cost of product extensions needed to keep ahead of competitors. Research has found that financial forecasting tools are not in favour of radical technology innovations with long horizons of evolution. Therefore, careful consideration is advised in relying on the net present value (NPV) estimations and on using a financial criterion only.

Strategic scoring

The decision grid (figure 6.5) is most useful in conducting a preliminary evaluation e.g. rating priorisation for inclusion on the roadmap[9]. For tough discussions on values or ideas that are brought to the table, and that generate doubt between parking and pursuing decision, the additional

Figure 6.5
Decision grid for strategic choice in design innovations options.

cc Simonse and Hultink, 2017[9].

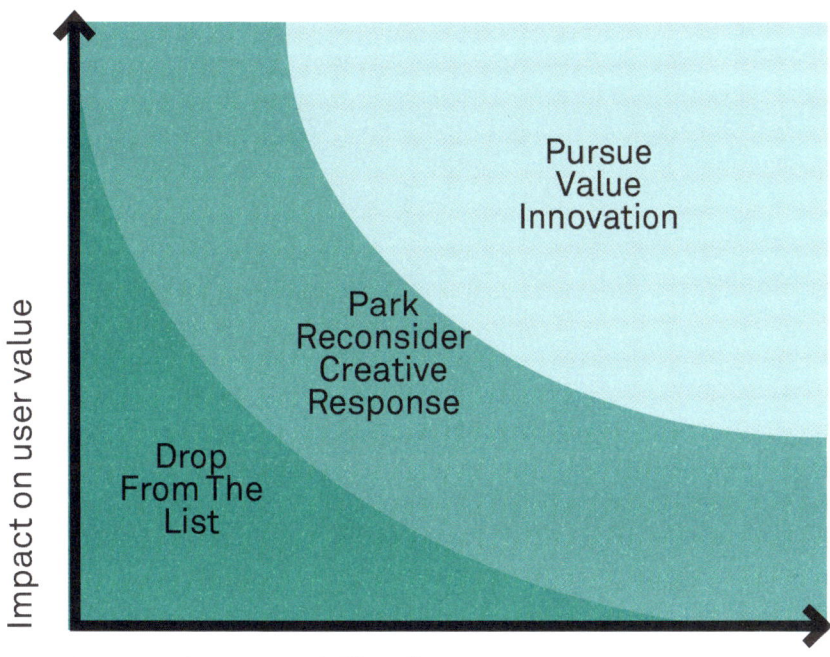

Impact on user value (y-axis)
Innovation capability fit (x-axis)

- Pursue Value Innovation
- Park Reconsider Creative Response
- Drop From The List

three core questions can be used.
Sometimes, roadmappers translate the questions into a scorings matrix. Then a scoring model of 1,3,5 or 1,5,9 is associated to the answer categories of low, significant, high. Then there are two ways to support an extended decision making on pursuing such design innovations. Either by allocating ownership of each strategic item to a professional role and asking this profession to make the final decision after intensive discussion or by making individual scoring forms and averaging the scores. Most roadmapping practitioners are in favour of the first way, as the benefit of the discussion is that knowledge is shared between the different organisations functions.

As the roadmapping session includes the making of decisions, of interest is the finding that fast decision makers use more, not less, information than slow decision makers[10]. Fast decision makers also develop more, not fewer, alternatives, and use a two-tiered advice process, interacting between tactical plans, such as roadmaps, and strategic decision making[10]. In general, fast decision making involves tactical plans that are created, with specialized input from experienced professionals. In dedication to strategic innovation decisions, design roadmaps offer such a tactical plan.

LAB ↗

Value mapping session

MATERIALS NEEDED:
→ note book
→ laptop
→ large sheets of paper
→ sticky notes
→ pens

1 BRAIN DUMP
On 'Changes in the world important into the future'. You can start by telling each other which change in the world you're interested in and be open to each other's reactions and interpretations. There is no wrong and right about the future. The goal here is to encourage the generation of a whole bunch of ideas on future changes and find those that resonate in the roadmapping team. In this way, you can learn a lot about how your team members think and what they might want out of a future vision.

2 TURN TRENDS INTO VALUE DRIVERS
'Trends and forces directly affecting our organisation'. To learn what's happening in the more direct environment, this step consists of sharing the outcome of the creative trend research and the technology scouting with a series of short presentations or talks. This sheds light on how users and organisations are, or could be responding to environmental changes, in search of new value wishes.

3 CREATE VISION STATEMENTS ON THE 'DESIRABLE FUTURE OF OUR ORGANISATION'
This step is to 'digest' these inputs by creating a vision of the organisation's most desirable future in subgroups of 3-5 people (see Lab ↗ Expression on a desirable future). The outcome is a series of agreed upon vision statements. These statements may include but are not limited to a definition of three to five value drivers.

4 VALUE MAPPING

Then the groups report out their vision statements to the whole group. Disagreement/differences will, of course, surface during this future visioning reporting. Create an agreed and disagreed list using the two questions that follow here to manage this part of the process:

→ Do you have any questions for clarification? As each group presents there are a number of terms and words that are discussed to avoid confusion and make their meaning plainly evident. Once everyone is clear, you can go on.

→ Is there anything up there in that desirable future of your unit that you could not live with or are not prepared to make happen? you can go on.

These two questions help rationalize, not resolve, conflict between groups so they can recognize and build common ground. (If they failed to ask these questions the groups can focus on what was different in their perspectives rather than what is common or similar). Once such conflict is rationalized, the roadmapping team is ready and committed to act in concert toward establishing innovative and creative relationships between the organisation and its environment.

5 DEVELOP A CREATIVE EXPRESSION OF THE FUTURE VISION FOR YOUR ORGANISATION

Then begin focusing on the areas of agreement — and integrating the work that the subgroups produced on the goals for the desirable organisational future is - which usually are a good stretch. By visually plotting out elements of envisioned product or service, you can learn a lot about the future vision. Not only will this help you refine what your future vision is, it can also reveal who will use it, where, and how. The idea of creating an image is to make something really rough as a way to help you think the vision through.

6 WRAP UP

Round off with commitment on the agreed upon desirable future. Then look back at the disagreed list to see if any of the items are still relevant — to the value drivers of the agreed upon desirable future. Any that still remain may be constraints to reaching the desired future. Move those for discussion in the next roadmapping session.

Prof. dr. HELEN PERKS on the subject of Cross-functional dialogues and Co-creation

LS Your research on Design in innovation shows a distinctive role of designers[11]. Regarding roadmapping, what do you consider as an important role for designers?

HP Well in that research we looked at multiple skills and activities that designers actually do and found three fairly distinctive roles of designers: first, a quite functional role in which designers stick to their specialist role and do what they are supposed to do, involving aesthetics, graphics etc. Then we saw a second role which is about teamwork and being really involved in a team. Whereas, the first role of the designer is a little bit siloed, in which one works as a sort of lone designer; the second role is where the design activities really would become part of a team project. The third role is where the designer becomes a leader of innovation, a process role in which (s)he would be involved in all different stages of the innovation process[11]. So for roadmapping, I would envisage that the team role would be important, because roadmapping is about working with others and getting buy-in from others, creating a cross-functional dialogue. Although this team role might seem a fairly established role, that is already considered really important in organisations, I have experienced nonetheless, that a lot of organisations still do not completely understand how a team works. They might have a team organised in principle, but then they still work according to 'you do your bit and I do my bit', and still the collaboration is about specifications and briefs. So the team role is getting over that, and having a cross-functional mind-set.

LS So the creative challenge is then also in helping the cross-functional team by mapping.

HP Well yeah, and then transferring that teamwork as a role into the mapping process. Because, roadmapping is very early on in the innovation process and when you are operating as a lone designer, then the chances are high that your ideas will not be understood by others until it is too late in the process. Therefore it also makes sense to organise creative dialogue sessions, in which there will be really quite a lot of trying on to find a common language. So the actual collaboration in roadmapping involves coming together and then you uncover that you are using different languages. It is often the case that the members of the roadmapping team have different perceptions and that they have different languages; they express themselves differently, they have different ways of thinking. You know, designers don't like being specific, often they like to be a bit hazy and creative, where the technology people they definitely want the 'I's dotted and the 'T's crossed, they want everything quite certain, whereas I think that designers are happily with ambiguity, and similarly, marketing people, they are very focused on the final outcome in the market, they all have a different content of interest. So with all these different backgrounds, different processes of working, then you come together in roadmapping, and have to try to find ways to encourage them, which is quite a challenge

LS When we go to the work that we have done. You identified shared visioning as an important concept in relation to the first mapping session[12]. Do you think it would be possible to co-create a vision with external stakeholders or is it better to do shared-visioning with internal stakeholders? What are your thoughts about that?

HP Co-creation is a very broad concept, and people interpret it in many ways. In the work I have done, co-creating is co-innovating together with customers[13]. Classically co-creation would be thought of in terms of involving external partners, partners outside the organisation: customers, users, other stakeholders such as suppliers and even competitors. Then it depends on the organisation, but I would think that if you can, co-creating with customers and users could be very powerful. Possibly at this very early stage, in terms of visioning, it would be very useful to think about what types of users would be appropriate for visioning. The established work on lead users found that they are very forward thinking, that they operate beyond the market trends and they are looking at solutions for advanced problems. So involving lead users in roadmapping could be very powerful. You have got to think about the setting. Classically lead users would be used in a business-to-business setting, so it would be organisations that have very advanced needs. The classic example is the car manufacturer who wants to develop a new breaking system, and invites a lead user from organisations in military aircraft design because they have very advanced needs for breaking systems. Military planes have to stop very quickly!. So, you have also got to look further for lead users, they are not in your existing customer base. For shared visioning, you need to be quite creative on what sorts of users you are involving because you need users beyond the existing marketplace. And when you envision in the consumer sphere, that means business to consumer, which obvious many organisations do, you still can find lead consumer users - those persons who are passionate about your products, and have started to adapt the products themselves and innovate around them.- The famous example here is Ducati, where motorbike riders are so passionate about their bikes that they are changing the bikes, changing the specifications. They are tinkering with it in their garage at the weekends and they have become part of the development teams. They behave like virtual teams in which they are working together. So what roadmappers also could do, is first to identify who their lead consumer users are and then send them out specifications for the new envisioned product.

LS So you also see in the makers-movement that people spend a lot of time on creative adaptions of products

HP And organisations are moving towards that way. If you only think of the increase of 3d printing and user toolkits, where they can configure and customize products to their own requirements. For co-creation this would mean you can identify people who like doing that and who do that a lot. The latest insights on this are that the people who were thinking about user configuration, who are using the user tool kits; these are the ones who are very loyal and passionate about the brand, they are highly involved. However, even when they become very active in helping develop and customize the product, new research has found that this does not move to purchase. Although they have self-designed these products, they still do not consider themselves as design experts, and in the end they rather prefer products from the designers, because those are the experts. This I sense as a kind of trend that is coming through in relation to user-innovation and co-creation. The trend of uncovering: 'who is the expert?'. The user may help the developers with all sorts of products. But they are not satisfied with the end result because they are not designers - they do not have the skills.

LS What could be the greatest benefits of co-creation in roadmapping?

HP Well, there is the view that you can get more creativity because you have expanded the scope, the space, of sources of creativity. Rather than be limited to just doing it with internal employees, who are probably going to be looking at previous products and might sometimes be quite biased towards their own experiences. Then, when you involve and co-create with customers, there is also the benefit of a closer connection and better fit to their needs and wants. In co-creating they know what their own behaviours, own needs and wants are; leading to a better fit with the customer requirements. Through co-creating together, you get richer insights to some of those needs and wants of which customers would struggle to articulate in traditional market research. As for example in a survey, that the questions are almost determining needs and do not give customers any scope for input or it is very banded. Whereas, if you are co-creating and, or use co-creating forums, you allow them the space. Or you can use ethnography where you observe customer's behaviours. I will always remember the case study work I have been involved with, with the shoe company Clarks. The designers in that company wanted to develop a new range of walking boots, and they actually went walking with customers in the mountains, observing and talking to them along the way, as they actually walked in their boots. In this way they would get their opinions which, if they would have asked them in a traditional way, maybe would not have provided those kind of rich insights. Another guy, also a designer at Clarks, went to his daughter's school and asked a load of kids to take of their shoes and put them back on again and he noticed how they squirmed to do so, and that hardly anyone bent down to undo the laces on their shoes. That triggered in him this idea of a self-closing magnetic mechanism for shoes. That sort of rich insights would else be very difficult to grasp with traditional market research. You also should consider the downsides of co-creation. It can be costly, time consuming, and involve customers who are not skilled designers. So they may come up with some ideas that are not feasible or not designable.

LS So roadmappers are also looking for differentiation, for the new target groups, the new values, to conceptualise new value proposition for. Have you seen market creation fuelled by co-creation?

HP Absolutely, you can co-create with your customers or you could be even working with other stakeholders as well and in that case particularly you can get all sorts of new insights for new markets. For instance you could co-create with a supplier who lets you know that they have this new component or technology, and share with you their 'what if' thoughts on developing an application and how that would open up a different market if you could reconfigure your product or bring in new resources. Or let's face it, co-creation in an alliance with competitors could be powerful as well. Classically, by merging of competences or technology capabilities you can co-create in fantastic new areas, and we often see that happen with new market creation. I saw an example of a microwave with a TV built into it. That is creating a new market for all those people who have got small kitchens; those who want to watch TV but they do not have the space for both a TV and a microwave, but when you combine those together and you get two companies let's say 'a Samsung' and 'a Panasonic' working together, then they may create a new market. These practices would primarily happen through co-creation and this is a very powerful mechanism to open up new markets that would be difficult to envision on your own.

> LS Also when organisations want to make use of trends, take for instance the IoT or Social media trends, for which they often do not have the capabilities in place yet, when you want to grasp and make use of that trend, then you need to collaborate with other organisations in the network.

HP Genuinely, I think organisations have to collaborate at an early stage as well. Look for instance at everything that is going on with mobile technology. Mobile companies are collaborating with digital consumer electronics companies and they are building and bringing out all sorts of fantastic new applications. Then with the new trend of the sharing economy, if you are sharing things there is a huge number of stakeholders involved; so beyond dyadic co-creating you may need to organises a whole network-setting, not just only the company working with a supplier or customer. You have to think quite creatively about that because there are lots of different stakeholders. I mean, at this moment I am doing research on value platforms. We say that it is very difficult to envision, right at the start of the innovation process, where that value is coming from when you are collaborating within a network for a new product or service idea. Particularly if you look at services, we found that value innovation really is an iterative process, in which incrementally you are trying to develop value around the initial idea by the network. This concept of 'value platform' is not the product or service, but a configuration of resources upon which various network members would add value. You need to understand their interpretation of value, and if you are quite open to that you can really come up with something very interesting and very innovative. Due to that openness the various stakeholders are more likely to buy in to that, during and at the end of the process. This is the same with roadmapping, because you are at the early start of the innovation process, gathering ideas and then you work on it by yourselves. We think that when you involve network partners to develop this value platform, the end result may be a product or service. But the idea is that the value platforms represent the value that everybody wants to see. The value platform is actually the product or service but as it is being developed, in moving on, from its early stage of figuring out the increasingly complexity and vagueness on 'What is the offering?'. This is very topical where the distinction between product and service is becoming blurred.

LS You also see that a lot in roadmaps. There is a shift from product roadmaps to service - and system roadmaps, as it is the service part where most of the innovation is envisioned, while the core product may just stay the same. Actually, the most promising opportunities of innovation are augmenting around the services. And that brings me to the last question, what do you recommend design roadmappers to do then, when they need to include services and service creation?

HP The shift from thinking about products is complex. It can create all sorts of perceptions if you solely define the value in innovations as 'products'. You need to put attention to the terminology that you are using in relation to creating a service. Because, by changing the terminology, it can help change the logic, the thinking and it can be more around values and allow the value platform to open up to the potential of co-creating, whether that is with the whole network or with each different organisation. By using our new terminology of value platforms, organisations can buy into that, while seeking to address what role are we playing, what do we value, what is important to us and how can we influence others in the network. That could be quite different for different members of the network. When you develop value platforms, the value is expected to be co-created in the innovation process and then you don't have to worry about the value proposition in the end; it sorts itself out because everybody has already bought in to it.

HELEN PERKS is Professor of Marketing at Nottingham University Business School, UK. Prior to academia she held senior positions with multinational groups throughout Europe, including Olivetti, the PA Consulting Group and the European Commission . She is Associate Editor (Europe) of the Journal of Product Innovation Management , academic Chair of the Product Development Management Association (PDMA) UK and Ireland and serves on the editorial board of a number of leading journals. She has over 100 conference and journal publications, with articles in leading journals such as JPIM, Industrial Marketing Management, R&D Management and Journal of Business Research among others.

↑
Stadia I, 2004

©Julie Merethu,
Artwork: ink and acrylic on canvas, 107 in. x 140 in. (271.78 cm x 355.6 cm). Courtesy the artist and SF MOMA

IN SUM

In this chapter, we have provided several guidelines to facilitate the mapping sessions. First to talk about the future values and share imaginations, gut feelings and beliefs. Second to map user values and new technologies to the products and services you envision and third to create and fine tune the design program roadmap for strategic review and decision making.

The three minimal critical dialogues we highlighted are:
- → Value mapping session
- → Idea mapping session
- → Pathway mapping session

Prior to the sessions, we advise to carry out design research activities on, for instance trends and technologies, or the strategic directions on time pacing. These activities in advance, provide the input for the dialogue sessions. For the follow-up and consolidation of the collective decisions we recommend a creative visualisation of the roadmap with artefacts of the future vision, the product roadmap or program roadmap.

To end, with respect to the facilitation of decision making across multiple innovation domains; in this chapter we showed that design roadmappers use a decision grid to prioritize and select new product and service innovations with the highest probability of overall success. The grid provides support and transparency in the tactical go/no-go decision making process.

---

1 Simonse, L.W.L., Hultink, E.J. & Buijs, J.A. (2015). Innovation roadmapping: Building concepts from practitioners' insights. Journal of Product Innovation Management, 32(6), 904-924.
2 Schein, E.H. (1993). On dialogue, culture, and organizational learning. Organizational dynamics, 22(2), 40-51.
3 Kim, E., Chung, J., Beckman, S. & Agogino, A.M. (2016). Design Roadmapping: A framework and case study on planning development of high-tech products in Silicon Valley. Journal of Mechanical Design, 138(10), 101-106.
4 Kawakita, J. (1982). The original KJ Method. Tokyo:Kawakita Reserach Institute.
5 Groenveld, P. (1997). Roadmapping integrates business and technology. Research-Technology Management, 40(5), 48 55.
6 Son, H., Kwon, Y., Park, S.C. & Lee, S. (2017). Using a design structure matrix to support technology roadmapping for product–service systems. Technology Analysis & Strategic Management, 1-14. Online available.
7 Phaal, R., Farrukh, C.J.P. & Probert, D.R. (2007). Strategic roadmapping: a workshop-based approach for identifying and exploring innovation issues and opportunities. Engineering Management Journal, 19(1), 3-12.
8 Day, G. S. (2007). Is it real? Can we win? Is it worth doing. Harvard business review, 85(12), 110-120.
9 Simonse, L.W.L. & Hultink, E.J. (2017). Design roadmapping: Managing transitions of the strategic life cycles. 24th Innovation and Product Development Management Conference (IPDMC), Reykjavik, Iceland, 11-13 June 2017.
10 Eisenhardt, K.M. (1989). Making fast strategic decisions in high-velocity environments. Academy of Management journal, 32(3), 543-576.
11 Perks, H., Cooper, R. & Jones, C.(2005). Characterizing the role of design in new product development: An empirically-derived taxonomy. Journal of Product Innovation Management, 22 (2), 111-127.
12 Simonse, L.W.L. & Perks, H. (2014). Cross-functional shifts in roadmapping: Sequence analysis of roadmapping practices at a large corporation. 21th International Product Development Management Conference (IPDMC), Limerick, Ireland, 15-17June 2014, Runner-up Best Conference Paper.
13 Perks, H., Gruber, T. & Edvardsson, B. (2012). Co-creation in radical service innovation: A systematic analysis of microlevel processes. Journal of Product Innovation Management, 29(6), 935-951.

VISUALISE ROADMAPS

207 VISUALISE ROADMAPS

HOW TO VISUALISE ROADMAPS

Due to its visual nature, a design roadmap is a powerful tool in facilitating communication between different stakeholder groups and across organisational silos. Yet, creating a visual roadmap is much more than mere graphic design, it involves both designing a map-structure that appropriately communicates the message of the future vision and its key innovations, and making the map visually appealing. Visualising a roadmap requires a great deal of logical thinking and creativity. Visualisation is certainly not something you do at the end, after the roadmapping process is complete. Visualisation starts at the very beginning by figuring out which structure will bring the clearest and most comprehensive overview.

In many organisations today, roadmaps are created by technical staff who have in-depth knowledge of technologies and project development but who often lack design skills. As a result, the maps they create lack communicative power, which limits adoption throughout the organisation. Applying a design approach to roadmapping offers the advantages of including visual communication power to a roadmap, hereby achieving the following:

→   Unravel complexity: design roadmaps make highly complex situations easy to understand;
→   Enable cross-functional decision making, by placing innovations in context: it is easier to prioritise projects and assign resources;
→   Better grasp the dynamics between markets, product and service offerings and technologies;
→   Highlight important events, such as strategic shifts, product launches and releases.

The ultimate goal of a roadmap visualisation is to make the innovation strategy self-explanatory in its actions.

This chapter provides you with insights and guidelines on the messages you want to get across to your target audiences, to visualise a roadmap covers both designing the structure and making it eye-catching. It requires a great deal of logical thinking and creativity. Visualisation is certainly not something you do at the end, after the roadmapping process is complete. Visualisation starts at the very beginning by figuring out which structure will bring the clearest and most comprehensive overview. The best roadmaps affect the viewer, are visually compelling and easy to comprehend.

Target audiences and communication needs

DEBORAH NAS

Creating a roadmap visualisation starts with identifying your target audience(s). Each target audience has their own needs regarding what they want to see and understand from your roadmap, which, in turn, helps you define the level of detail required. The roadmap design shapes the layers of information that members of your audience will use to make informed decisions. In our experience, there are three target audiences with different levels of decision-making influence[1]. The first audience to

TARGETS AUDIENCES with decision influence are senior management, product programmers and network stakeholders.

persuade on aligning their priorities to future innovations is the senior business and R&D management. Second are the people responsible for the ongoing coordination and programming of design innovation projects – product and program managers. Third are people from outside the company, whom we call 'network stakeholders'. These can be the organisation's suppliers and co-creation partners, people involved in strategic partnership activities, consortia, NGOs or representing the general public, journalists and innovation experts.

↘ Senior management

Senior management has the power to impact the organisation and the projects it undertakes. They devise the corporate strategy and usually decide when a new business proposition will be introduced to the market. They also decide how many of the company's resources will be invested in innovation. At start-ups, senior managers often participate in designing the actual roadmaps, but at corporate organisations, they become the first audience to convince. They are a key target audience because when they approve the roadmap, they agree to the future vision and the development pathways you and your team have defined. Although their lens of interpretation may vary per organisation and context, they often use roadmaps to:

- → Identify strategic targets for development programs;
- → Create foresight on what possible trends create compelling opportunities;
- → Understand what course of action a future vision can lead to in terms of new products and services;
- → Identify key technologies that offer competitive advantage in the long run;
- → Recognise how strategy, new product and service offerings and innovation projects relate to each other.

When presenting your design roadmap to senior management (see an example in figure 7.1), the challenge is to gain acceptance by complying with their communication needs. Your roadmap design should provide them with a good overview of the process and facilitate their strategic decision making. Its purpose is to enable them to decide when to launch new value propositions; prioritise, budget and allocate resources and align priorities and decisions across functions and departments, including design, marketing and technology. Link these different aspects together, and balance a strategic view with an overall system level of detail.

↘ Product & program management

The people responsible for the ongoing coordination and programming of innovation projects are the organisation's program managers, product managers, design managers, marketing managers and so on. They initiate and review design and innovation projects. They influence the prioritisation of design innovation projects and the direction of projects they are involved in, and operate on a more tactical level than senior management. Some of them might be part of the roadmapping team; others only need to be kept in the loop regarding future innovation plans. Product and program managers tend to use roadmaps to:

- → Identify new projects that can achieve strategic innovation targets;
- → Commit to the timing of launching new products, services and/or features, and accepting the responsibility to make this happen;
- → Understand the sequence of successive innovations per product line;
- → Identify how new product and service offerings relate to innovation efforts across functions and departments, and use this to decide on (future) projects to realize them;
- → Comprehend how partner and supplier efforts contribute to making the proposed innovations real.

For serving this audience your roadmap design should tie the organisation's innovation strategy and future vision together with the proposed innovation projects, and hence the roadmap for them should provide more details about new product and service launch timing and follow-up releases, plus the timing and resources involved. The roadmap's design and presentation for them should help them to program the timing of the projects to realise the proposed design innovations, assign the appropriate project leaders and team members to projects; and initiate projects with a project budget, delivery deadlines, and a project brief. For the project and program management, the roadmap is likely to be part of quarterly or (semi)yearly strategy review sessions, and thus will need to be updated accordingly. Therefore, for keeping the roadmap up to date, make sure your roadmap design is made in a format that non-designer stakeholders can work with. This is the most challenging need and, unfortunately, it often limits a bit your design freedom since most stakeholders do not use graphic design programs.

↘ Network stakeholders

External network stakeholders have diverse needs that make designing a roadmap tailored to their viewpoint a real challenge. Here are some examples of network stakeholders and their needs:

- → Suppliers and co-creation partners, who want to learn about the organisation's innovation strategy since it directly impacts their innovation activities;
- → Shareholders, who want to know how the company will be making money in the future – this might include

DESIGN ROADMAP
Port of Amsterdam

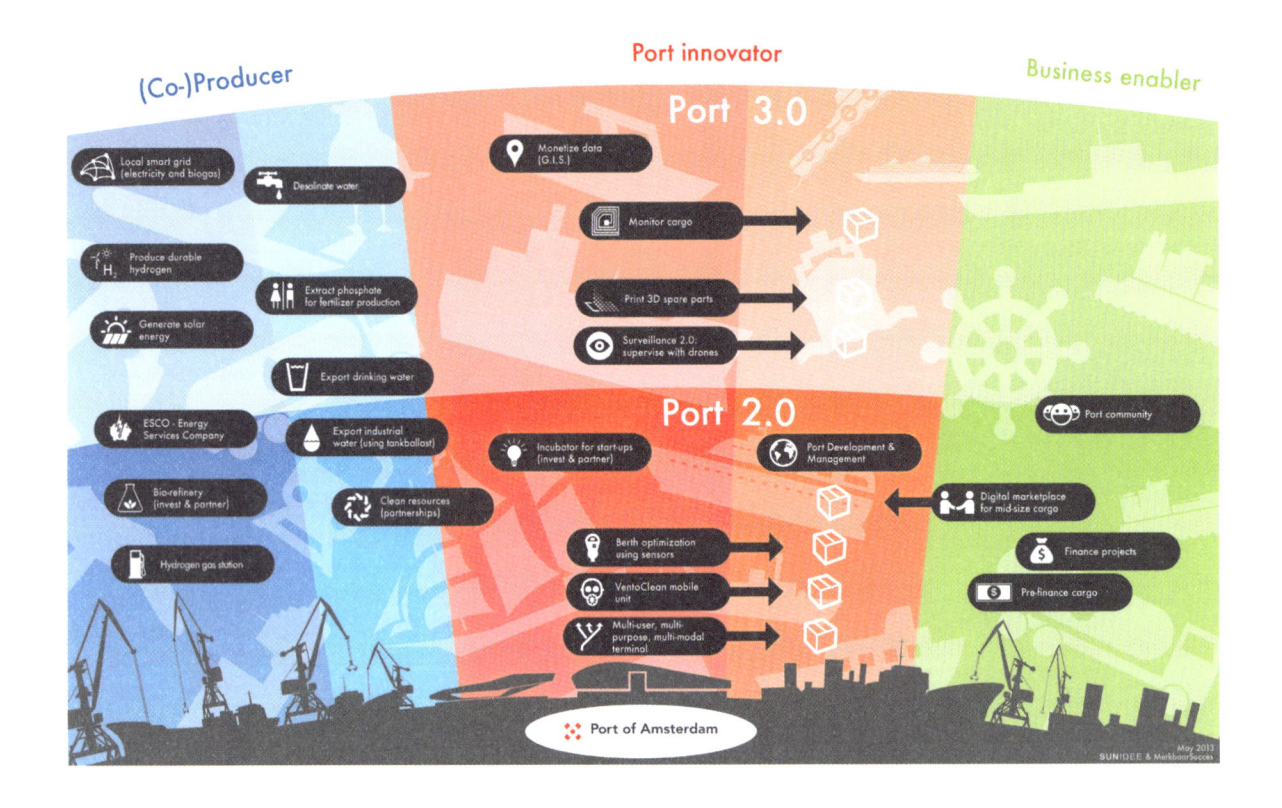

Figure 7.1
Design Roadmap
Port of Amsterdam

© SUNIDEE & MerkbaarSucces, 2013.

- → breakthrough innovations, new business models, geographical expansion, etc.;
- → Business-to-business customers, who want to feel confident that the company is pursuing continuous innovation, and will rely on their customer-supplier relationship;
- → Legislators, NGOs and representatives of the general public, such as journalists, who are interested in learning how the company impacts the world on a societal, environmental and economic level.

For these audiences, your roadmap design should be extremely visually appealing and self-explanatory. In some situations, it might also be used as a marketing/PR tool. Typically, the roadmaps you create for network stakeholders should exclude confidential information that is shared with the internal audiences. Therefore, edit the roadmap's content carefully – together with the roadmap's internal stakeholders, consider upfront which confidential information must be excluded. Remember: communicating with external audiences amounts to public disclosure of the organisation's innovation plans. It is a very sensitive document, since its content should be agreed by the stakeholders with the commitment to make the roadmap come true. The only exceptions to this rule are roadmaps that are made as part of a future exploration study, showing what could be possible and purely intended to inspire people and inform strategy making.

How to's

Getting different stakeholder audiences to commit to your roadmap is not an easy job; there is often the "not invented here" hurdle. Here are some guidelines in this regard[1]:

- → Identify and interview the target audience members you choose to work with carefully. Ask senior stakeholders whom to assign particular tasks, which will ensure that future roadmapping team members are those trusted for their professional expertise and decision-making skill.
- → Keep 1-2 senior managers closely involved in the roadmapping process. They can for instance give a short speech when the roadmapping team first gets together. They can also take part in progress update meetings, where they can give their feedback about intermediary results.
- → Find a balance between including team members with strong social networks within the organisation and more 'hardcore' content specialists – consider at the start of the roadmapping process to interview stakeholders to understand their needs and expectations but also to find out whom to invite for the roadmapping team.

Designing a roadmap template

DEBORAH NAS

Designing a roadmap template can guide you in achieving your communication goals. The challenge is to find an easy-to-understand structure that can incorporate the often large amounts of data and relationships between the data. Roadmaps are information-dense, your challenge is to make that information easy to absorb for different target groups. The format and schematic overlay you choose should help viewers to see things in context and quickly identify key topics relevant to them[2].

Which format is best?

This is a tricky question. If you are a designer, you probably have good graphic design skills, so your ambition would be to create something amazing that everybody loves and uses. Designing a roadmap template implies deciding on the number of roadmaps you will have to produce[3]. If it is possible to achieve all your communication goals using a single, easy to read, visually appealing roadmap – great! Otherwise, you will need to design entirely different roadmaps for different audiences[4]. Still, try to limit the number of roadmaps as much as possible to prevent confusion within the organisation and keep version control manageable.

Choosing the right medium – a poster, talk, video or website, for example – entails understanding what you want to achieve and how people will use it. For instance, the roadmap you present to senior managers will primarily be used to communicate the company's future vision and inspire them to get on board with it. The message will most likely stay the same for a couple of years, which justifies the budget required for a multimedia production. The same is true for the more commercial roadmap intended for the network partners, as it leaves out the most confidential information.

Programmers have a particularly restricting and challenging requirement. They tend to use roadmaps as living documents – also known as dynamic documents because they are always in use – and thus the roadmap needs to be updated on a regular basis (often every quarter, or half year) and each innovation team is responsible for keeping their part up to date. Updated versions are typically shared throughout the organisation in digital form – as PDFs, or as part of PowerPoint presentations – and sometimes physically, in the form of printed posters. Updating the tactical roadmap is part of a structured innovation management process. The updates are programmed at quarterly or

VISUALISE ROADMAPS

Structure, format and connect elements in the
ROADMAP TEMPLATE.

longer intervals and, typically, product or program managers do not make changes to the roadmap themselves.

Creating a template for your roadmap

→ TIMELINE FOR TELLING THE STORY

Important in shaping the template is the reference to the timeline that makes it easy for different managers to find a starting point to read and explore the roadmap. A product manager will start in the area describing the offering, and from there explore relationships to grasp the bigger picture; a technical developer will start with the technology building blocks, and so on. In many roadmap designs, the timeline runs from left to right, although you might like to come up with alternative solutions such as a curve . The timeline element of the roadmap supports you to tell the innovation story about the pathways that lead to the future vision. The timeline element directs viewers' attention toward the horizons and innovations that lead to the future vision. It has a beginning, a transition in the middle and an ending[2]. The story told by the timeline is often partitioned into three horizons, which together communicate the logical sequence of the envisioned business evolution. And by further explanation, the parallel lines of different types of design innovations can be highlighted.

→ DEDICATED SPACE AND SHAPES

Other visual elements that determine the roadmap template are the spaces, shapes, or icons, used to indicate key innovation elements[3]. Different shapes and spatial relationships can be used to express values of target groups, key technologies, new versions of products and services, new business model creation and adoption, and so on. Below, we present three basic roadmap templates to help you get started on

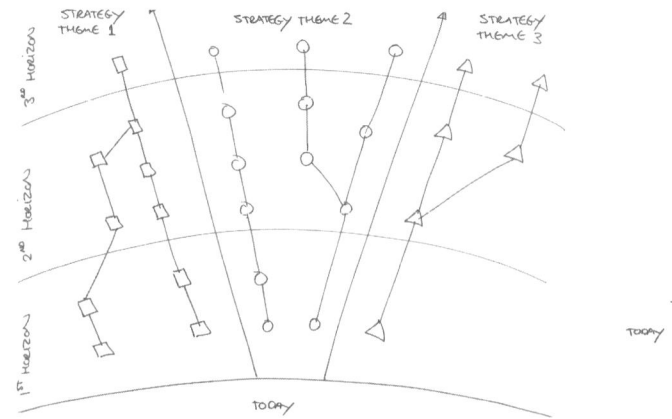

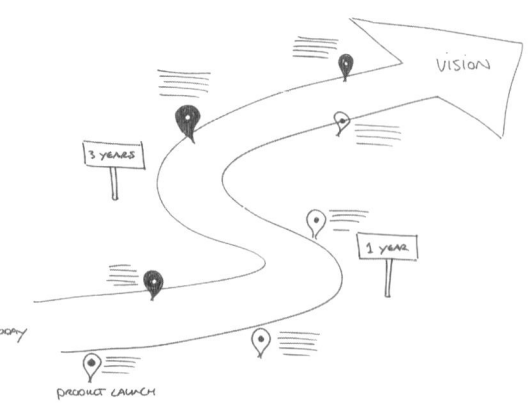

Figure 7.2
Sketch of an example roadmap layout for senior management

Figure 7.4
Sketch of an example roadmap layout for network stakeholders

cc Nas, 2017.

Figure 7.3
Sketch of an example roadmap layout for project & program management.

cc Nas, 2017.

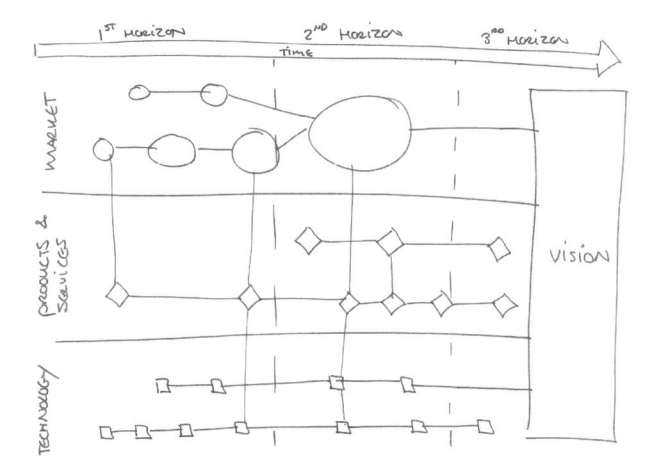

your visualisations. As you gain more experience, you will develop your own set of templates, each of which optimally serves the needs of your target audiences.

→ DEPICT THE PRODUCT, SERVICE, USER VALUE, MARKET AND TECHNOLOGY RELATIONS
To help different stakeholders align their various innovation efforts, and gain a wider sense of the direction and timing of future innovations, it is important to clearly illustrate innovation pathways and the relationships between each innovation element. For instance, it is worth distinctively visualising the cohesive aspect module additions in a product line, and highlighting the connections between new user values and product visions in the tactical roadmap. Bear in mind that the fine tuning of product/market/technology content is an iterative process that takes place during the roadmapping process.

The example roadmap template in figure 7.2 situates the present day at the bottom. Three horizons extend radially outward from "today" to emphasise that the scope and variety of the company's business activities will expand over time. The vertical lines, dotted with symbols to indicate various project milestones, trace the parallel development of three strategic innovation themes. This roadmap template was designed to communicate the future vision strategy to senior management in the organisation.

The roadmap in figure 7.3 is a tactical roadmap template, targeted to project and program management. Its layout features 'swimming lanes' that together depict the three main sections related to innovation: user value/market, product/service and technology. The timeline, with its three horizons, runs from left to right in this template. Since project and program managers are responsible for the ongoing coordination and programming of innovation projects, the symbols and their spatial alignment are used to indicate new product and service launches, follow-up releases, timing and resource allocation.

The roadmap layout in figure 7.4 uses a more high-level, creative approach. It visualises innovations over time. This roadmap layout is useful when communicating innovation plans to network stakeholders. Since network partners are external to the firm, and often include the wider public, there is not too much detail in the roadmap, and the timing is merely indicative.

Creating compelling visualisations

DEBORAH NAS

After the content and layout for the roadmap have been defined, your next task is to use your graphic design skills to turn the roadmap into an effective communication tool. This means designing something that is beautiful, logical and inspiring. Make consistent use of colours, icons and fonts, and limit the amount of text. The roadmap should also communicate well as an element in a presentation slide or printed out on paper. Your goal is to make the target audience feel interested, happy, and upbeat when reading your roadmap. Good roadmap visualisations have the quality of presenting information in a visually compelling and easy to process manner. It is of great service to decision making if your roadmap is easy to comprehend, graphically consistent and affects the viewer.

↘ Easy to comprehend

As you might have experienced yourself, 'a picture is worth a thousand words'. Several studies have confirmed that adding visuals leads to better outcomes (compared to text only). Recent research on visual communication suggests that, at the most basic level, visuals contribute to the impact of information by increasing viewer attention, comprehension and recall[5]. Thus, content in text combined with visuals is more likely to be encoded and comprehended than is information presented in text only. Ultimately, the roadmap should be self-explanatory. This can be achieved when the target audiences perceive that the roadmap provides information they need in an effective way.

↘ Graphic consistency

There are no strict rules for designing a graphically consistent roadmap; however, the following tips and tricks provide you with some useful guidance:
→ Less is more. If you want your audience to easily understand your roadmap, don't overload them with information. Choose

VISUALISE ROADMAPS

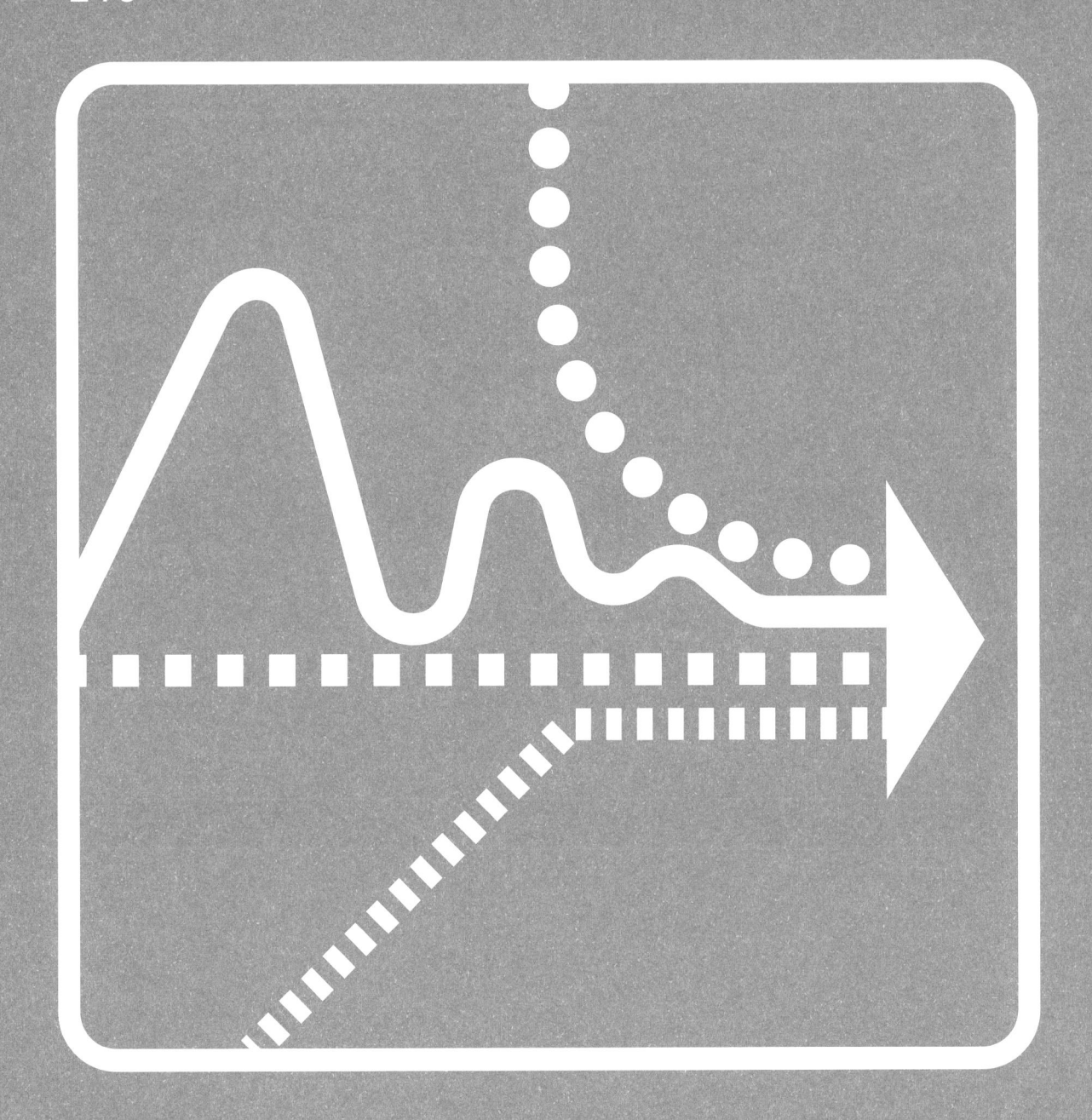

Getting everyone on the SAME PAGE.

wisely which information is important enough to be part of your roadmap; don't clutter it with information it doesn't need. Also, if your roadmap contains the same information in several places, this is usually a sign that changing its layout will result in a more simplified view.

→ Clearly depict the timeline. Make sure that the 'today' and 'future vision' portions of the timeline are plainly visible – they act as visual anchors where viewers begin to read the roadmap. Help your audience to easily understand how the innovation program will develop over time by extending the timeline horizontally or vertically.

→ Use colours sparingly. It can be tempting to colour-code everything in your roadmap, but unfortunately, this often leads to visual clutter. Ideally, limit yourself to 1-3 colours. Work with colour variants – different types of blue, for example – or use a small colour set and apply different levels of transparency. Keep in mind that people who are colour blind should also be able to read the roadmap, so do try to avoid red-green combinations.

→ Limit text, and use 'clean' fonts. Short paragraphs of text can best be placed around the edges of your roadmap; limit text in the roadmap itself to a few key expressions. Choose a font type and size that suits the medium of your roadmap – in print, some font sizes appear smaller than on screen. Also, pay attention to font contrast (the difference between the brightness of the background and the font).

→ Use icons appropriately. Icons are a great way to communicate a message if used well. Regarding look and feel, the icons need to work with the overall design of your roadmap. Don't overpopulate your roadmap with icons; think wisely about location, size and how many different icons to use. If icons have a widely accepted meaning – like location pins – make sure to use them in the same way to avoid confusing your readers. Size circles by surface area, not diameter when using circles to depict relative importance. Limit the number of different icons, and include a legend if needed.

→ Keep it clean – use 'flat design'. There is already so much going on in a roadmap that 3D images are often just too much and are distracting.

↘ Affect the viewer

Getting your message across is about more than content alone. People simply may not accept or pay attention to visual graphics they do not like. Researchers have defined affective criteria of personal attitudes towards visuals by asking people to what extent (on a scale of 1 to 5) they thought a visualisation seemed not cool/cool, boring/interesting, unpleasant/pleasant, unappealing/appealing, not likable/likable, unexciting/exciting, and unattractive/attractive[5]. Part of the challenge to affect the viewers is finding ways to set a positive emotional tone. The emotional tone determines the extent to which the roadmap grabs your target audience's attention, and makes them feel happy and upbeat. When senior managers or programmers experience such positive feelings, the roadmap is typically more readily adopted, more often shared with colleagues, and managers show more commitment to making it a reality.

Take an iterative approach

Our experience has shown that you will need numerous iterations to finalize your template design. It's not just iterating on the graphic elements of the design, but also on the structure and content. A fruitful way of collecting feedback is to show the roadmap template to someone who was not involved in creating it – without explanation. Ask them to think out loud, and explain what they see and understand and how it makes them feel. Ask open questions to gain a better understanding. Based on your learnings, create an improved version and repeat the procedure a couple of times until you are content with your efforts and proud of the result.

Figures 1.4 and 7.5 show two example roadmaps designed for Quby, a European market leader in smart thermostats . In the first chapter of this book, figure 1.4 shows a strategic roadmap for the management team. Below, figure 7.5 presents the tactical roadmap for the audience of project programmers. For the strategic roadmap in figure 1.4 the team adopted a playful design, visualising the three horizons as three circles. Only the most important information is incorporated into each horizon: key societal and technology trends, hardware and software highlights, key partners and a simple visualisation of the innovative product-service system. The colour scheme is limited to different types of blue, which goes well with Quby's logo and is 'easy on the eyes'. The designers found the right balance between minimizing the amount of information and stimulating the readers' desire to learn more. The roadmap also illustrates that a timeline doesn't have to be a straight line. The designers creatively sketched a unique format to get the message across.

The tactical roadmap in figure 7.5 depicts the three horizons in far greater detail. It coherently links market, product, technology and resource developments. The designers chose a simple, linear timeline, so as to more easily interlink all the elements. The schematic overhead layout is readily apparent to viewers, helping them to navigate the roadmap and find areas of interest. Icons are used to draw attention to the 'swimming lanes' (market, product, technology, resources); and the product section and future vision sections – elements that the team wanted to highlight – really stand out.

DESIGN ROADMAP

Tactical roadmap for QUBY

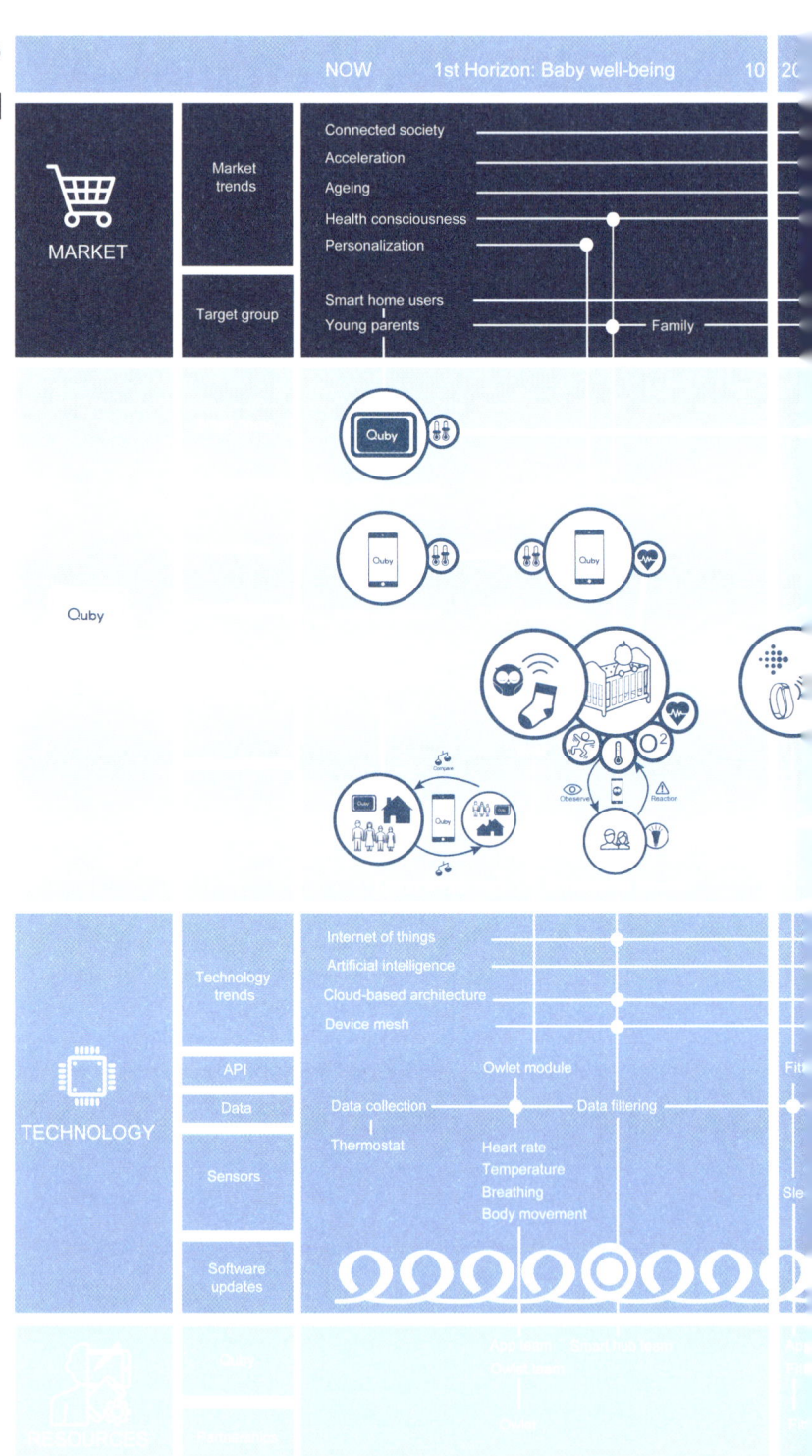

→
Figure 7.5
Tactical roadmap for QUBY

cc Pepijn van Dalen, Luuk Roos & Zoë Dankfort. QUBY Project Report, Design Roadmapping Master Course. Faculty Industrial Design Engineering Delft University of Technology.

Please note that the design roadmap is created for Quby by Strategic Product Design Master students, and therefore do not reflect Quby's actual strategy.

223

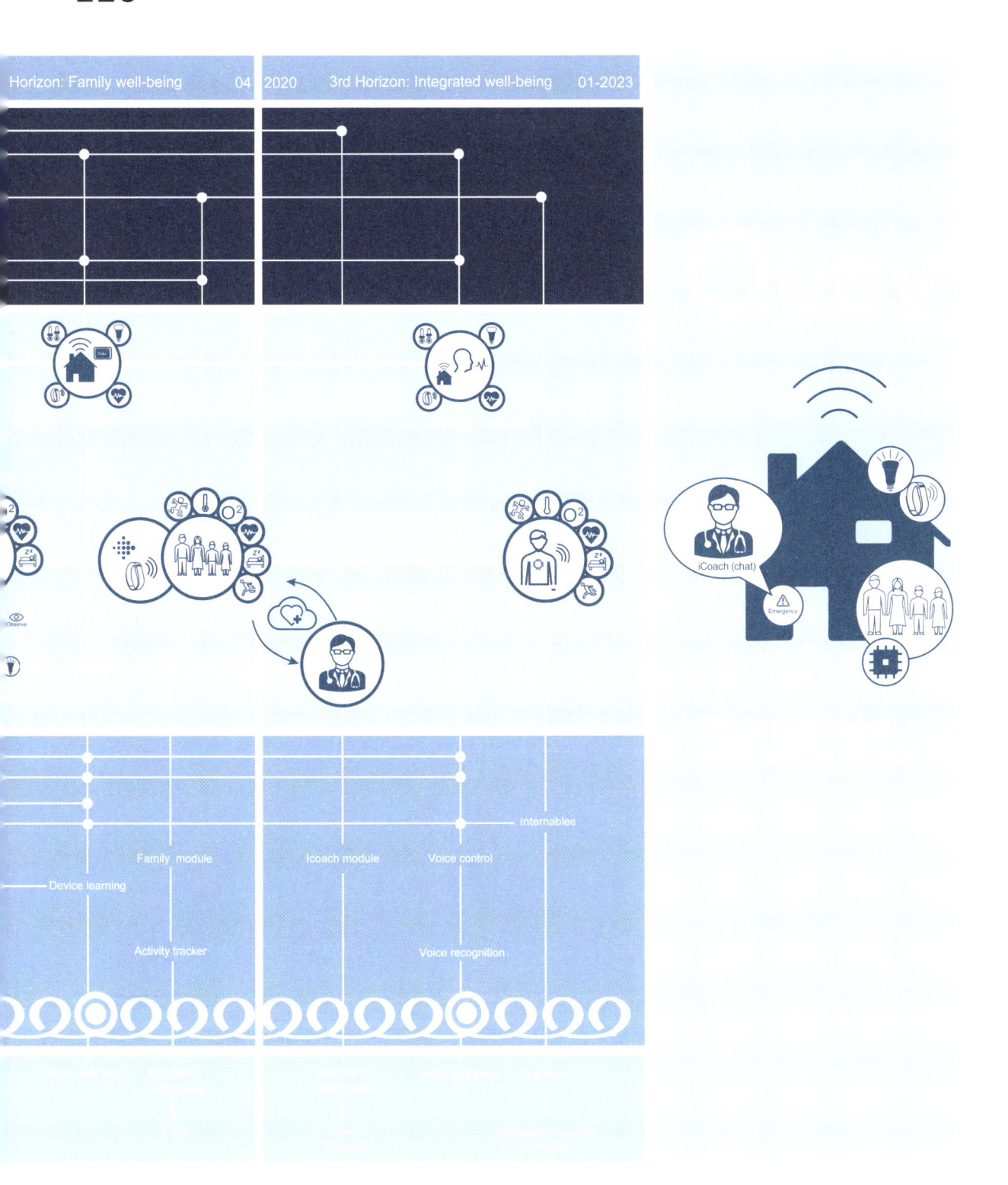

VISUALISE ROADMAPS

Horizon: Family well-being 04 | 2020 3rd Horizon: Integrated well-being 01-2023

Family module
Device learning
Icoach module
Voice control
Internables
Activity tracker
Voice recognition

STATE LIBRARY CASE

Service roadmap
SASHA ABRAM, SABINA POPIN AND BIANCA MEDIATI

↘ Maps and mapping: manifestations of co-design

Mapping is a formidable tool to navigate complex landscapes and to visualise, understand, and guide us on the literal and metaphorical journeys of designing in the 21st century[7]. At Meld Studios, we use maps both as an object-in-use and as a way of-imagining . Maps play enduring roles in maintaining the strategic direction of an organisation long after our consultancy has ended. Mapping allows conversation and reflection across hierarchies and departments, around the pain-points in the delivery of a service and, inevitably, within the organisation itself.

> "It becomes accessible to groups at various levels of engagement and enables opening of conversations via the pen and hand.

It is a creative methodological process, which simultaneously acts as participant capacity building and research gathering[8]." Mapping provides opportunities for employees to build an ontological understanding of their own roles within a company as a "way of coding a reality we 'know' but can never really see for ourselves[7]," in many cases giving staff a new perspective. Following is a case study of the State Library of Victoria (SLV).

↘ Co-designing the future of the State Library of Victoria

The SLV engaged Meld Studios to redesign the library's services. Over a twelve-week period, an integrated participatory design team – three designers from Meld Studios and two staff from the library– sought to understand the current state of the library's service delivery and the opportunities for a service model redesign that would enable the organisation to take on a new public role in the digital future. This meant that the co-design project had an additional component of skills transfer: training the internal project team to develop the capabilities to take ownership of the vision going forward, and giving them the confidence to lead the next phases of work so that all library staff could be given the opportunity to shape their own future.

N.B.
Case description based on the paper presented at the ServDes conference 2016[6]. Reworked by the authors.

Figure 7.6

Affinity Mapping, combined with sketches of the physical space, was used to structure, contextualise and externalise the insights for the whole team to access.
Over 2 weeks of research, insights were collected.

© MELD STUDIOS

Affinity and current state mapping

The project began with two weeks of intensive research (the 'Understand' phase): shadowing, observing and interviewing customers as they interacted with the library and its services. The project team also talked to library staff and senior managers to help identify strategic themes for the new direction of the library. To process the scale and complexity of the research insights, the team chose to externalise and share their research using affinity mapping: "We mapped our collective insights and did our analysis and synthesis upon the walls, rather than trading documents created by each individual (see figure 7.6). We captured our research observations and insights on post-it notes then clustered them into groups to see the bigger picture and identify patterns. We did our thinking outside of ourselves and made sense of what we were hearing and seeing as a team.[9]"

Maps visualise the complexity and interconnectedness of end-to-end services – their relational geographies – through vignettes of service scenarios for both front - and back end of the house. In current state mapping, low fidelity maps are created – either hand-sketched or designed with a deliberatively naive visual style – to encourage participation and to emphasise the iterative nature of the design concepts in progress. The 'unfinished' approach mitigates intimidation for non-designer stakeholders and foregrounds a focus on interacting with the concepts within the maps as they are still being formulated. The findings from the research were recorded in four current state journey maps covering the following themes: interacting with the physical space, information and collection access, the library as a place to work and community engagement and programming. The maps were made widely available for the library staff to review and critique. In many cases, this was the first time staff had been exposed to co-design methodologies.

Participation in the mapping process allowed for an introduction into the democratic process of participatory design, drawing upon the Scandinavian tradition, which advocates that:

"people who are affected by a decision or event should have an opportunity to influence it."

HUSSAIN, SANDERS & STEINERT[11]

↘ Future state map: service roadmap

After concept testing through prototyping and playback, we 'articulate' potential future service possibilities through design, in a future state map. The future state map is a highly visual and refined artifact, designed to be a lasting reference point for organisational change. Compared to the current state map, the future state map is a an introduction into the democratic process of participatory design, (see figure 7.7 and 7.8) . Higher-fidelity map, with more refined concepts, that is used towards the end of a project to deliver a vision of a possible future-state scenario and the roadmap for achieving this institutional change for our client.
The Future state map utilises the richness gathered from research insights and generative concept workshops, references challenges in the

→
Figure 7.7
The Future State service roadmap.
Here presented to senior stakeholders.

© MELD STUDIOS

It is currently being used to implement the Library's service strategy over the next 5 years.

current state reality and converts them into opportunities for the future. It is through this forecasting that new offerings, products, services, spaces and behaviours can be explored and eventually implemented. In order to 'Realise' a service redesign, we used the future state map as a service roadmap to facilitate a plan for enacting the future state. The roadmap encapsulated a single vision of the new library service model. It was 3 metres long, and designed to be viewed on a wall and shared with people, in contrast to a report that is read and absorbed individually. In this way, the design of the map actively encouraged discussion and collaboration.

The new library vision captured in the future state roadmap has sparked over 30 individual projects, to be undertaken over the next three to four years (the 'Realise' phase), some of which have already been implemented. "At a distance, the map gives a broad overview of the future service. At the macro level it shows an integrated service model that clearly places the collections at the heart of everything we do, with services built around the needs of the customers instead of around our internal workflows or the physical layout of the building, a deliberate decision in order to future proof the model should we make changes to the configuration of the building[10]."

For the embedded team, additional value came from designing, prototyping and testing the future concepts for 12 weeks within the library. This created opportunities for the people who would need to implement and live them to manage anxieties about change in situ, identify advocates to help lead the change, identify who had the power to block change and hear the challenges verbatim. This all significantly contributed to the success of the project.

The future state map of the state library now sits on the CEOs wall as a daily reminder of their 5-year vision. Maps and mapping played an essential role in the redesign of the service model for the SLV. The ability for maps, to display and decipher a significant quantity of information was fundamental to the success of the project at every stage. The large scale and flat format of the maps allowed openings for democratic discussions and collective feedback. The fidelity and visual accessibility of the maps created opportunities for library staff to add their expert insights without inhibition. The social practice of mapping encouraged both advocates and critics for change, to make their voices heard. The collaborative learning of mapping skills by the library staff has empowered them to implement some of that change.

At MELD STUDIOS, we use maps both as an object-in-use and as a way of-imagining . Maps play enduring roles in maintaining the strategic direction of an organisation long after our consultancy has ended. Mapping is a formidable tool to navigate complex landscapes and to visualise, understand, and guide us on the literal and metaphorical journeys of designing in the 21st century.

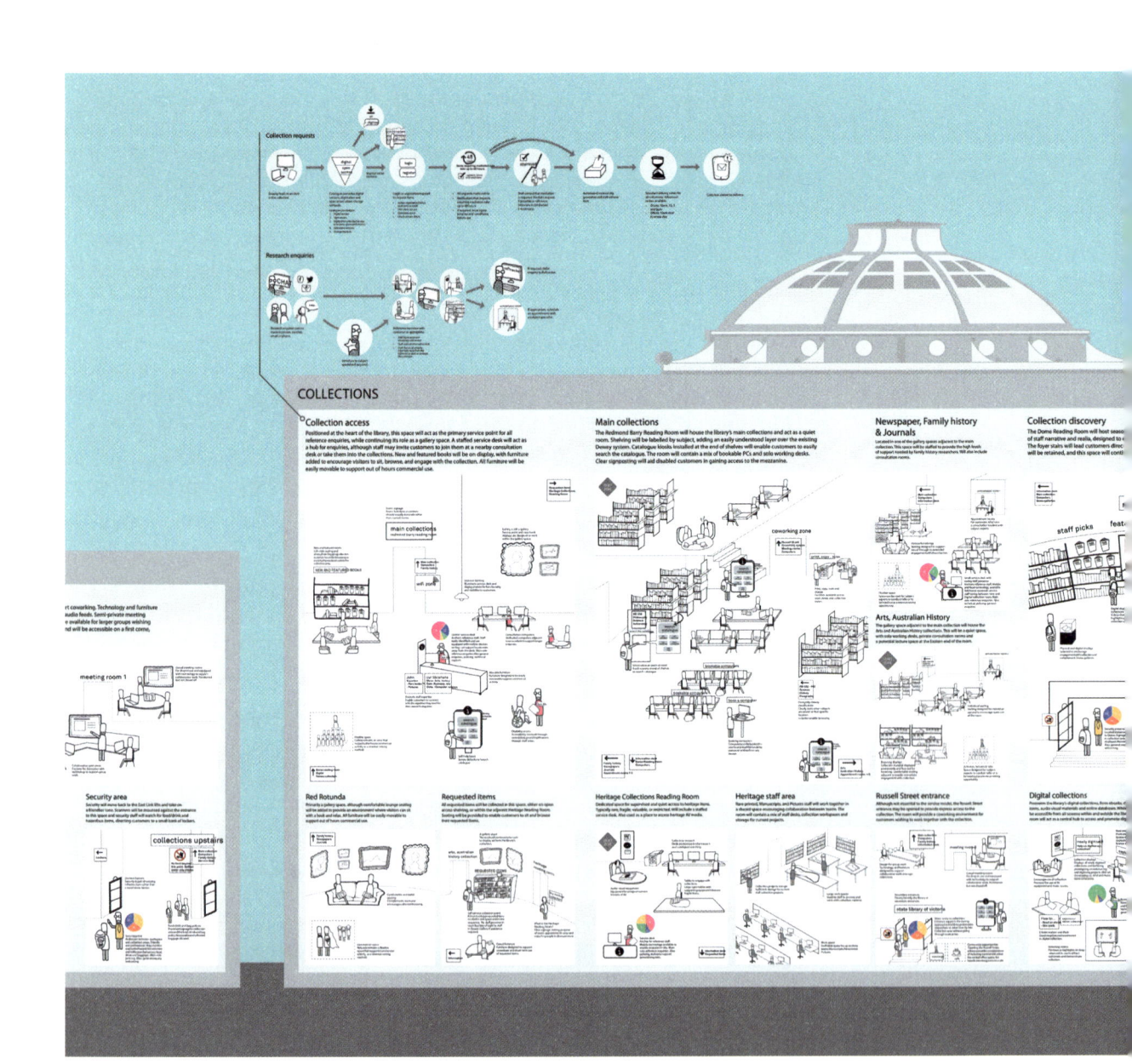

FUTURE STATE SERVICE ROADMAP
State Library of Victoria

229

VISUALISE ROADMAPS

↓ →
Figure 7.8
Future State service roadmap of the State Library of Victoria, Australia.

© MELD STUDIOS

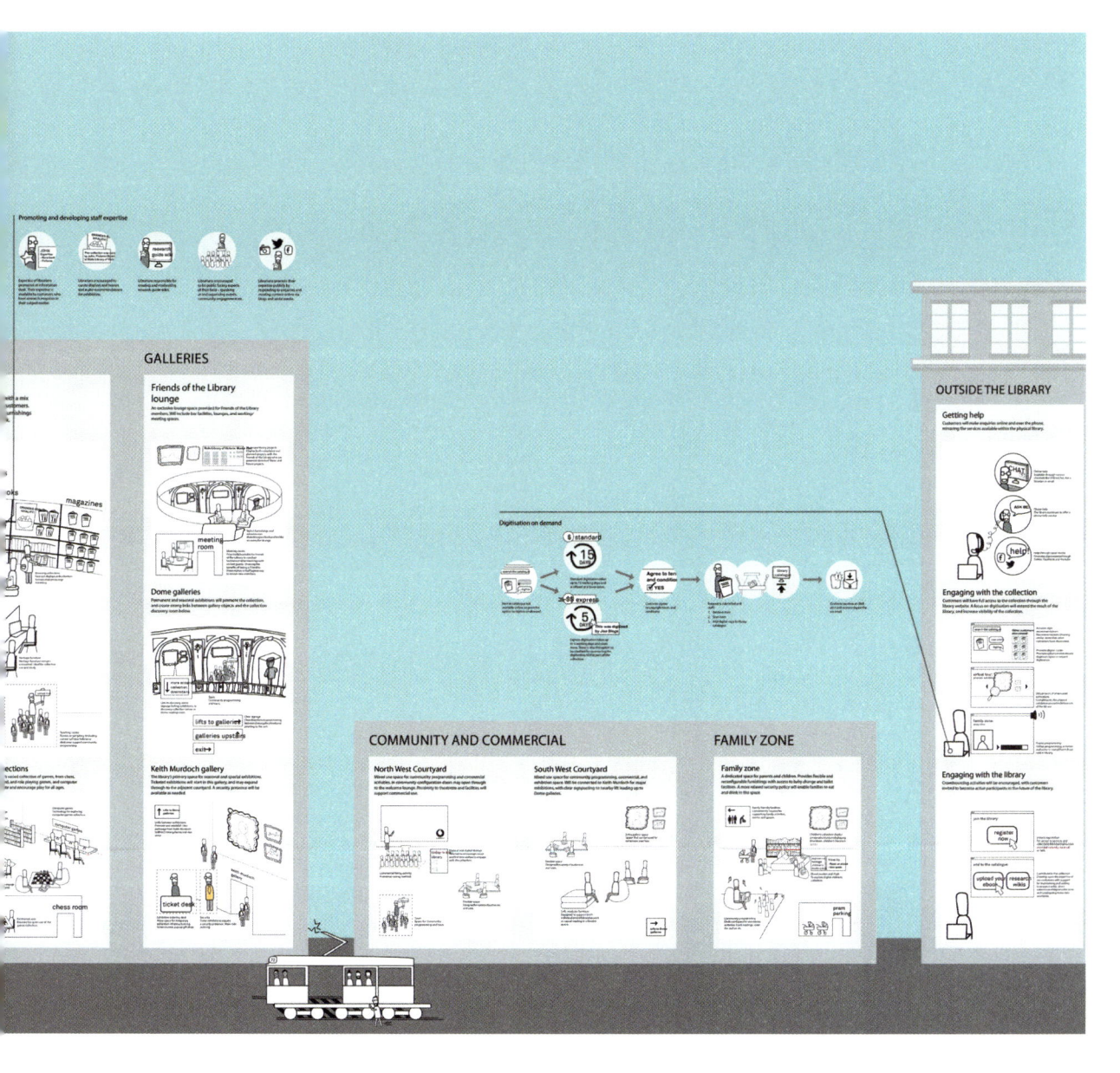

LAB ↗

Create your roadmap template

MATERIALS NEEDED:
→ note book
→ laptop
→ large sheets of paper
→ sticky notes
→ pens

1 Identify which target audience(s) you're designing for: senior management, project programmers and network partners. And within these target audience(s), ask yourself which types of stakeholders you want to address and what their needs are with regards to the roadmap. In other words: identify what you need to communicate to whom.

2 Decide on the number of roadmaps- Use one map if you think it can achieve all your communication goals. Otherwise, design different roadmaps for different (groups of) stakeholders. Try to limit the number of roadmaps as much as possible to prevent confusion within the organisation and keep version control manageable (a roadmap is a living document).

3 Choose the format for your roadmap- Understand how often the roadmap will be updated, who will be involved, and through what medium the roadmap will be shared. Based on these requirements, choose the best format for your situation – a printed poster, digital presentation, video, website, etc.

4 Structure and direction- Design the timeline and the template format for your roadmap: this is the core structure of the map. Start with some clean sheets of paper and begin sketching. Iterate towards a clear and sound structure that effectively communicates the strategic direction, the future narrative.

5 Content, relations and key events- Map all roadmapping elements in your template. A simple way to do this is to take a large piece of paper, sketch out the structure of the roadmap and then use sticky notes to position the elements which constitute the actual content of your roadmap. Identify relationships and draw lines to connect different roadmap elements. Iterate your way towards a simple, well-structured and logical layout.

6 Graphic Design-Use your graphic design skills to turn the roadmap into an effective communication tool. This means designing something that is elegant and inspiring. Make consistent use of colours, icons and fonts and limit the amount of text. The map should also communicate well as a slide in a presentation or printed out on paper. Your goal is to make the target audience feel interested, happy, and upbeat when reading your roadmap.

7 You will likely have to go through between 3-5 iterations in this step to finalise your template design. You are not just reiterating the graphic design of the template, but the structure and content as well. The best way to approach this is to have 2-hour in-depth discussions with a small team to discuss the roadmap and identify improvements. You create a new version, have another in-depth discussion, create a new version . . . until you and the key stakeholders are happy and proud of the result.

Prof. dr. ROBERTO VERGANTI on the subject of Meaningful innovation and Design Roadmapping

LS My first question is about your previous book Design-Driven Innovation[12]. In it, you highlight the distinctive role of designers, their capabilities and design methods. So my question is: What do you consider the most important role designers play in terms of innovation strategy?

RV You know, the thing is – as regards innovation, designers make sense of things valued by users. They take huge amounts of information and make use of their imagination to really design what life is all about. Designers have the capabilities to take the users' perspective, and really understand what they want and what is important to them. This is surely different from engineering-driven innovation, where business managers see things from the perspective of shareholders and engineers from the perspective of machines. For designers, innovation is all about meaningfulness. Then, because creating value for the user is of course important for the innovation strategy, designers have definitely the capabilities to shape. Not only to shape the solution, but certainly on the strategy level of innovation, to shape the direction.

LS My next question is about your research[13]. You have identified the importance of creating meaningful innovations. In roadmapping, the creation of a future vision is what gives the roadmap its direction. To what extent do you consider it is possible to map meaningful innovations?

RV First of all, you cannot really design a 'meaningful' thing, because meaning is ascribed by users. The actual meaning is decided upon by the users. But in that respect, what you can do is design a platform of meaning built out of user insights, through which you can envision differences in various users' values. Although it is not always possible to find that these depend on meaningful design. And, to a large extent, this is the situation right at the moment – there are so many innovations, so many ideas, that we've got overcrowded

SCIENCE INTERVIEW

with innovations, often the continuations of innovations from previous times. Generally, I have the impression that we all feel a little bit fatigued with innovations. There are so many new things around, all these quants of innovations, and they are not necessarily meaningful, although it is possible to design meaningful innovations. As a designer, I think, we need to move away from quantity and go for the quality of innovation. We need to shift from the quantity of ideas and how many times we can renew innovations. We need to keep dreaming about the revolutions that design can bring. At best, you try to only design things that have value in them. These meaningful innovations you can map on the roadmap. The most important thing is that you choose to map not quantity, but quality. That you are critical on the innovations, you map only those new innovations that will be meaningful in the future.

    LS And do you think it is possible to *envision* meaningful platforms then?

RV Of course, there is a big problem here, because over time in a changing world, what is meaningful changes as well. If you focus on solutions, you will inevitably end up solving problems that meanwhile have become meaningless. When I talk about meaningfulness, I am talking about the now. All these things we live with now and that have changed our way of living did not all happen because we thought they would be possible now. We might have had a couple of ideas, take for instance, AirBnB – one little bit that was imagined in advance was about property sharing, but all other ideas relating to the meaningful experience of AirBnB services today have not been imagined in the way that is possible now. I am also a little concerned about people who can look into the future and attempt to think about it in a quite linear way. Because then it ends up with new things that only build on what we know already today. So, in reality, with meaningful innovation, the future is today, in the sense that, since the future did not happen, because no one made them, I am not sure you can say something meaningful will happen 'in the future'. I think there is the now; there are many things that did not happen but that could have happened – and can happen if you envision them. Sure things can become meaningful some day in the future, but that is a matter of someone loving the thing or someone not loving it.

    LS Part of the roadmapping process is the simultaneously mapping of different types of innovations. Do you distinguish between different types of meaningful innovation, or do certain groups of users determine different types of meaningful innovations?

RV   As an Italian, the way I ascribe meaning to things might practically be different from what you give meaning. Meaning is a reflection of what occurs in our personal lives, which also depends on the socio-cultural model and the adaption to the evolution of it. Then there is symbolic meaning, which is a social good that you give meaning when you have significant use in relation to others. And then you have emotional meaning, which is about how something creates a different mindset.

LS   So part of being involved in roadmapping also stimulates thinking about taking on the job of 'creative lead' or becoming the visionary leader driving the innovation strategy. What is your advice on becoming a visionary leader?

RV   You can read my new book 'Overcrowded[14].' I propose, exactly the opposite of what is needed to become a visionary leader -not someone who is creating a vision outside-in. First, to design new meaningful directions, we need to start with ourselves, not with outsiders. We need to work from the inside out. Solutions can be borrowed from the outside, since they enable us to achieve a target, but the target- the direction -has to come from within us. The design of solutions moves from the outside in. For example, user-centred innovation advises designers to start by going out and observing how users use existing products; open innovation advises designers to engage with outsiders and have them propose new ideas; and when it comes to our leadership contribution, we are advised to be visionary and 'think outside the box'. Instead, the design of a meaningful direction works the other way around: from the inside out. We need to start from ourselves, from what we find meaningful. Rather than jumping outside of the box, we change the box from the inside. Second, we need criticism instead of ideation. Beyond creative problem solving that is built on the quantity of ideation, the innovation of meaningful directions requires that designers become skilled at the art of criticism. Because when you propose a new direction, you start from a tiny inkling of an idea, and you discuss a hypothesis together. And often, such initial proposals are blurry and quite vague. You just started to carve out a sense of direction, whose value and implications are still unclear — not only to others, even more so to ourselves. Criticism enables us to dig deeper, confront our insights with the insights of others, and find a new, more powerful purpose that lies underneath. It also prevents us from getting stuck in old directions and helps us to get rid of a past that might not be meaningful anymore. In an overcrowded world, the value of designers is not to provide more ideas, but to provide developmental feedback.

LS So, should envisioning a meaningful direction always be a team activity, or can it be an activity carried out by an individual designer?

RV Envisioning is something you do individually first, then in pairs, then in a circle. This is to focus your individual creativity and the creativity of others towards a shared purpose. And, yes, I think it is important to bring other people in; however, you need to create innovation from the inside out. To agree on a few directions with each other, and then on what follows, that you are all open to interact with what happens. The process begins with a vision and proceeds through developmental criticism, first from a sparring partner and then from a circle of radical thinkers, then from external experts and interpreters, and only then from users. In this way, inside-out, during the envisioning activity, the direction will become more meaningful. And at the same time you will also become more sure about what the meaningful innovation is not. There are a few things essential about this: a vision is subject to criticism and it does not come from outsiders but from people's unique interpretations.

LS For my last question, I'd like to know what you recommend design roadmappers to do, in creating more meaningful roadmaps?

RV My recommendation would be, first, to take the people's perspective and see that people change a lot and that there are dynamics in their needs. Then also see that there are some basic needs that do not change – they only change in meaning over time. Second, stay connected with deep underlying values by reflecting on your own needs and values, and spar with others on this in a critical and constructive way. Then you will uncover the reasons why we need things, and move your innovation one level higher. Before you design a new solution, you first have to design the new direction.

ROBERTO VERGANTI is Professor of Leadership and Innovation at Politecnico di Milano, where he teaches in the School of Management and the School of Design. He directs Leadin' Lab. Roberto's research focuses on how leaders and organisations create innovations that people love. He explores how to generate radically new visions, and make those visions come real. His studies lie at the intersection between strategy, design and technology management. Roberto has issued more than 150 articles. His research has been awarded the Compasso d'Oro. He is the author of the books "Design-Driven Innovation"and "Overcrowded". Roberto has served as advisor at a wide variety of manufacturing and service firms including Ferrari, Ducati, Procter & Gamble, Unilever, Gucci, Nestlé, Samsung, 3Ms, Zappos, Microsoft, IBM, Vodafone.

DESIGN ROADMAP

Strategic roadmap for QUBY Smart Thermostat

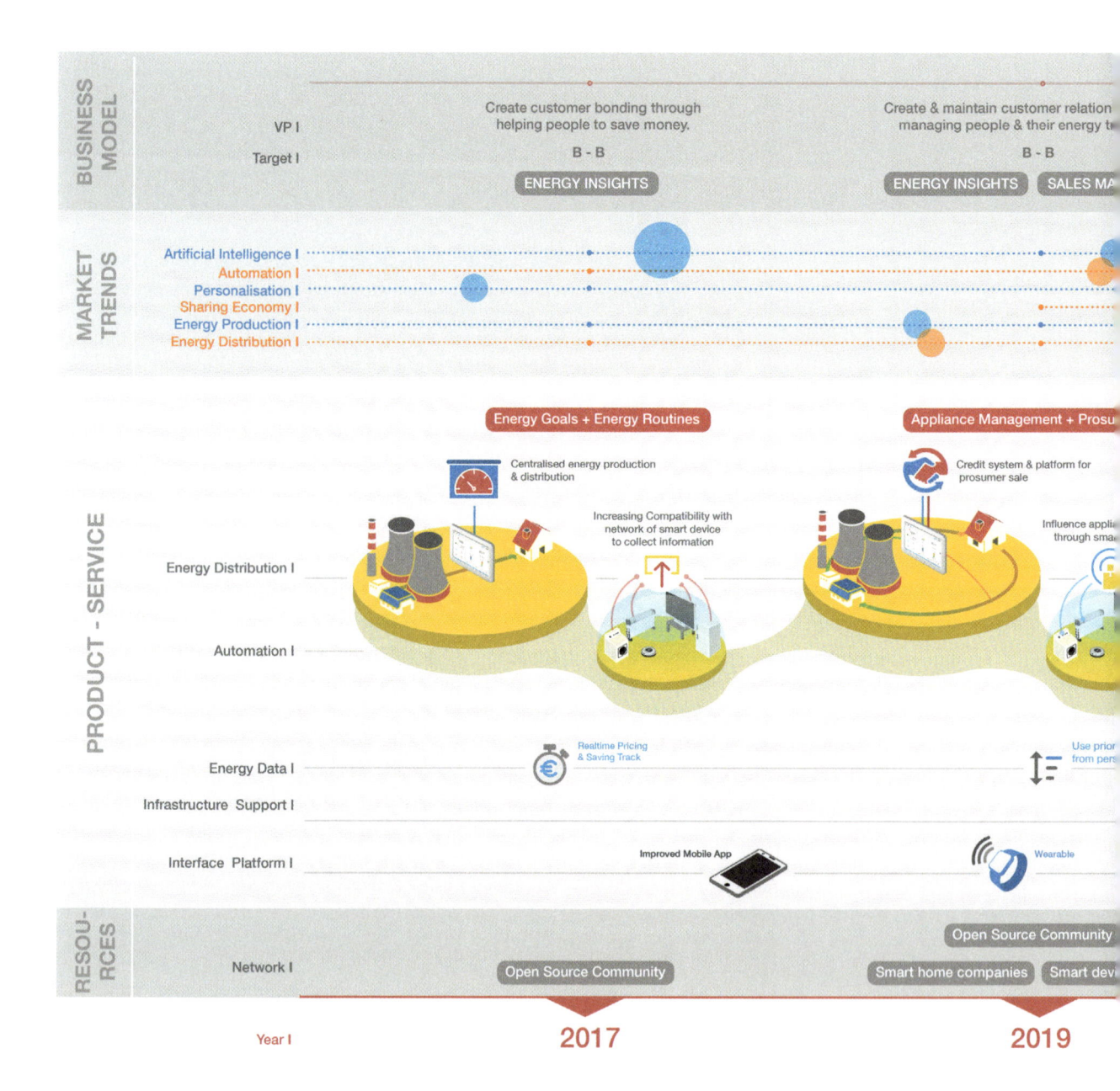

BUSINESS MODEL

VP I: Create customer bonding through helping people to save money. | Create & maintain customer relation managing people & their energy t...
Target I: B - B | B - B

ENERGY INSIGHTS | ENERGY INSIGHTS | SALES MA...

MARKET TRENDS

- Artificial Intelligence I
- Automation I
- Personalisation I
- Sharing Economy I
- Energy Production I
- Energy Distribution I

PRODUCT - SERVICE

Energy Goals + Energy Routines | Appliance Management + Prosu...

- Energy Distribution I — Centralised energy production & distribution | Credit system & platform for prosumer sale
- Increasing Compatibility with network of smart device to collect information | Influence applia... through sma...
- Automation I
- Energy Data I — Realtime Pricing & Saving Track | Use prior... from pers...
- Infrastructure Support I
- Interface Platform I — Improved Mobile App | Wearable

RESOURCES

Network I | Open Source Community | Open Source Community / Smart home companies / Smart dev...

Year I: 2017 | 2019

237

VISUALISE ROADMAPS

Please note that the design roadmap is created for Quby by Strategic Product Design Master students, and therefore do not reflect Quby's actual strategy.

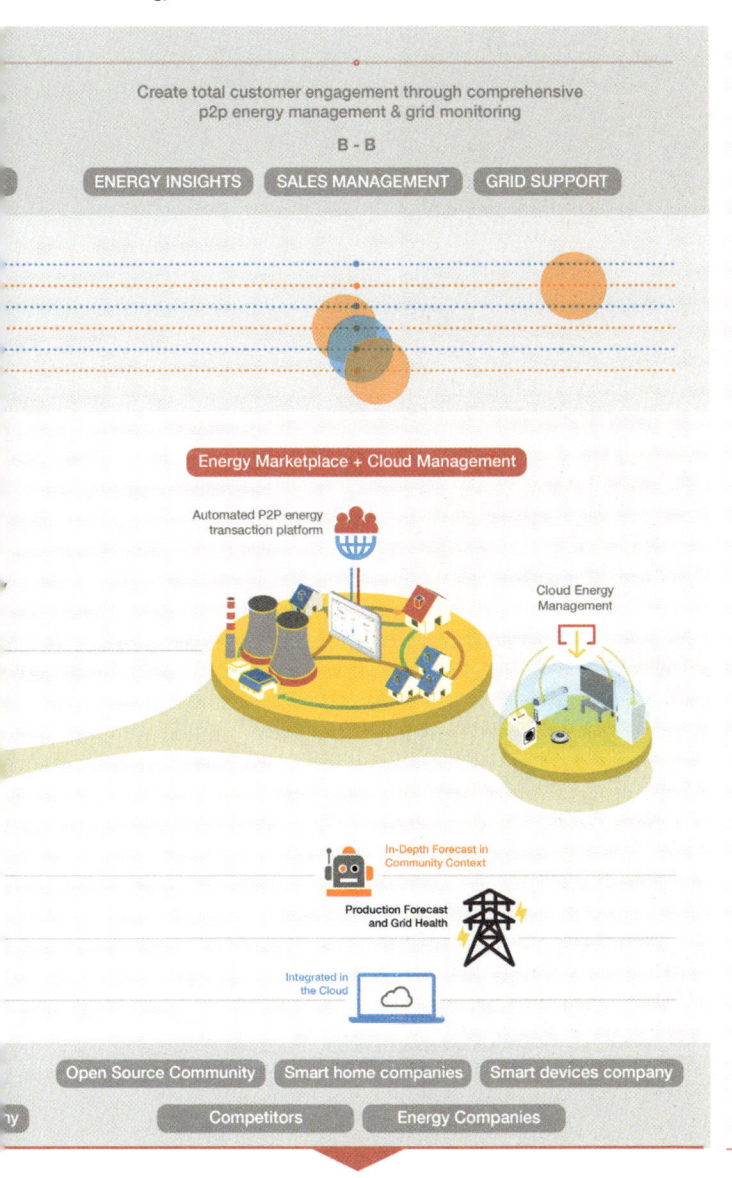

↓
Strategic roadmap for QUBY

cc Zhang Ziyi, Yee Jek Khaw & Roël Tibosch QUBY Project Report, Design Roadmapping Master Course. Faculty Industrial Design Engineering Delft University of Technology.

Create total customer engagement through comprehensive p2p energy management & grid monitoring

B - B

ENERGY INSIGHTS SALES MANAGEMENT GRID SUPPORT

Energy Marketplace + Cloud Management

Automated P2P energy transaction platform

Cloud Energy Management

In-Depth Forecast in Community Context

Production Forecast and Grid Health

Integrated in the Cloud

Open Source Community Smart home companies Smart devices company

Competitors Energy Companies

2021

Vision for Quby 2030

Smart energy solutions provider anywhere and anytime

Everything connected

Energy in the Cloud

2030

238 DESIGN ROADMAPPING

←↖ *Starry Night*, the Van Gogh-Roosegaarde bicycle path.

© Daan Roosegaarde and Studio Roosegaarde with Heijmans.

The smart bicycle path is made of thousands glowing stones inspired by 'Starry Night'. The path charges at daytime and glows at night. Here innovation and cultural heritage are combined in the city of Nuenen NL, the place where Van Gogh lived in 1883.

Smart Highway, the interactive and sustainable roads.

IN SUM

In this chapter, we have provided several guidelines to help you contend with the challenge of visualising a roadmap, and creating a roadmap template.

First by identifying the communication needs of your target audiences. They will use the information contained in the roadmap in different ways related to their decision making responsibilities. For example, making choices about strategic directions and high-level project priorities requires less detailed information than deciding how to assign project leaders and team members to specific projects.

Second, we recommend that you sketch a roadmap template at the beginning of the roadmapping process and then iterate further upon it as the process unfolds, by using it as a living document during the creative dialogue sessions.

A third recommendation is that you try to be graphically consistent and engaging with your roadmap designs. The guidelines for this are:

→ Use graphic elements and colours in consistent ways;
→ Incorporate icons to increase communicability;
→ Include a positive emotional tone to make your target audience feel interested, happy, and upbeat;
→ Iterate on your design until you are happy with it, and feel proud of the result.

To get started we have provided several examples and templates, and showcased an inspiring visualisation project carried out by Meld Studios. Ultimately, we encourage you to develop your own signature roadmap style across a variety of media.

---

1. Create Innovation Roadmaps. Step by step approach for your Innovation strategy. http://sunidee.com/clear-growth-strategy/ accessed July 2017.
2. Simonse, L.W. L., Buijs, J.A. & Hultink, E. J. (2012). Roadmap grounded as' visual portray': Reflecting on an artifact and metaphor. 28th EGOS: European Group for Organisation Studies Colloqium-SWG 09, Helsinki, Finland, 5-7 July, 2012.
3. Kerr, C. & Phaal, R. (2015). Visualizing roadmaps: A design-driven approach. Research-Technology Management, 58(4), 45-54.
4. Phaal, R. & Muller, G. (2009). An architectural framework for roadmapping: Towards visual strategy. Technological Forecasting and Social Change, 76(1), 39–49.
5. Borkin, M. A., Bylinskii, Z., Kim, N. W., Bainbridge, C. M., Yeh, C. S., Borkin, D. & Oliva, A. (2016). Beyond memorability: Visualization recognition and recall. IEEE transactions on visualization and computer graphics, 22(1), 519-528.
6. Abram, S., Popin, S. & Mediati, B. (2016). Current states: Mapping relational geographies in service design. 5th ServDes Conference, No.125, 586-594. Copenhagen, Denmark, 24-26 May 2016.
7. Hadlaw, J. (2003). The London underground map: Imagining modern time and space. Design Issues, 19(1), 25-35.
8. Schultz, E. & Barnett, B. (2015). Cognitive redirective mapping: Designing futures that challenge anthropocentrism. Nordic Design Conference, Stockholm, Sweden, 10-11 September 2015.
9. Gagarin, D. (2014). Co-designing with the State Library of Victoria: How we did it, and why it paves the way for change. http://www.meldstudios.com/au/2014/06/10/co-designing-state-library-victoria-itpaves-change, accessed January 2016.
10. Hyde, J., Conyers, B. & Flynn, B. (2015). Journey maps and customer hacks: Redesigning services at the State Library Victoria, Synergy 13(1). http://www.slav.vic.edu/au/synergy/volume-13-number-1-2015/perspectives-local-/491-journey-maps-and-customer-hacks-redesigning-service-sat-the-state-library-victoria.html, accessed January 2016.
11. Hussain, S., Sanders, E.B.N. & Steinert, M. (2012). Participatory design with marginalized people in developing countries: Challenges and opportunities experienced in a field study in Cambodia. International Journal of Design, 6(2), 91-109.
12. Verganti, R. (2009). Design driven innovation: changing the rules of competition by radically innovating what things mean. Boston, MA: Harvard Business Press.
13. Salerno, M., Landoni, P. & Verganti, R. (2008). Designing foresight studies for nanoscience and nanotechnology (NST) future developments. Technological Forecasting and Social Change, 75(8), 1202-1223.
14. Verganti, R. (2017). Overcrowded: Designing meaningful products in a world awash with ideas. Boston, MA: MIT Press.

VISUALISE ROADMAPS

DR. IR. LIANNE W.L. SIMONSE

Lianne is passionate about innovation, design and strategy and its wonderful combination. She loves art, cars, and likes to travel. Her research interest are on design roadmapping, future visioning, creative trend research, qualitative research techniques, organisation design, socio-technical system design and business model design for digital services.

Dr.ir. Lianne Simonse conducts a scientific research track on design roadmapping and has presented papers, including an award winning, at design conferences of the Design Society, the Design Management Institute, ServDes, and on innovation management conferences of the Product Development and Management Association, Academy of Management and European Group of Organisational Studies. Lianne's scientific work is published in journals such as Design Issues, Journal of Product Innovation Management, Journal of Medical Internet Research and International Journal of Technology Intelligence and Planning. Based on this comprehensive work and her experiences with roadmapping in industry, this guidebook on design roadmapping has been written.

FEATURED CONTRIBUTIONS

ENA VOÛTE is professor of practice and Dean of the Industrial Design Engineering Faculty of Delft University of Technology.

FLAVIO MANZONI is Senior Vice President, Head of Design at FERRARI, Italy.

GLORIA BARCZAK is professor of Marketing at Northeastern University, Boston, MA, US, and editor in chief of the Journal of Product Innovation Management.

RICARDO MEJIA SARMIENTO is Design & Social Innovation consultant and PhD fellow at Delft University of Technology.

ANNA NOYENS is a Strategic Product- & Service Designer and earlier Chief Product Officer of PEERBY, Amsterdam.

SUSAN REID is professor of Marketing, Bishop's University, Sherbrooke, QC, Canada.

PAUL HILKENS is Vice President Materials & Device Technology Development at Océ-Technologies BV, a Canon company, NL.

ERIK JAN HULTINK is professor of New Product Marketing at Delft University of Technology.

NIYA STOIMENOVA & DIRK SNELDERS are a Strategic Product Design MSc and professor at Delft University of Technology.

PASCAL LE MASSON is professor Design Theory at Mines ParisTech —Paris Sciences et Lettres Research University, France.

HELEN PERKS is professor Marketing at Nottingham University, UK and associate editor of the Journal of Product Innovation Management.

DEBORAH NAS is professor of practice Strategic Design for Technology-based Innovation at Delft University of Technology, and co-founder of the strategic design studio SUNIDEE in Amsterdam.

SASHA ABRAM, SABINA POPIN & BIANCA MEDIATI are service designers at MELD STUDIOS, Melbourne & Sydney, Australia.

ROBERTO VERGANTI is professor of Leadership and Innovation at Politecnico and author of Design driven innovation and Overcrowded.

THE AUTHOR
Lianne Simonse teaches Design Roadmapping at the Industrial Design Engineering Faculty of Delft University of Technology in The Netherlands. She holds a MSc. and PhD. in innovation management and has combined her academic positions with 20 years of professional experience in industry. Lianne has executed and led several roadmapping projects and worked with different corporate organisations and start-up ventures on roadmaps. Dr. Simonse conducts a scientific research track on design roadmapping. Based on this comprehensive work and the experiences of how-to roadmap this guidebook on design roadmapping has been written.

THE ART DIRECTOR
Barbara Iwanicka is graphic designer and art director, based in Amsterdam. She attended Cooper Union in New York and Gerrit Rietveld Academy in Amsterdam. Barbara's art works derives from the intersection of art, architecture and design, focusing mainly on printed matter. Her work has been exposed in the Stedelijk Museum in Amsterdam, Printing Museum in Tokyo, and 41 Cooper Gallery in New York. She has been awarded with the Best Dutch Book Designs of 2014 and 2015 and was a finalist of several design and architecture competitions.

THE TEXT EDITOR
Jianne Whelton is a professional book editor, based in Melbourne, Australia. Jianne has over 15 years of experience teaching, consulting, translating and editing. She works with academics, policy advocates, research practitioners and professionals from a variety of domains.